REVISED EDITION

HOME PLACE

Essays on Ecology

STAN ROWE

FOREWORD BY WES JACKSON

The first edition of *Home Place* was published in 1990 and reprinted in 1992, under the ISBN 0-920897-78-9.

National Library of Canada Cataloguing in Publication

Rowe, J. S. (John Stanley), 1918-
Home place : essays on ecology / Stan Rowe. — 2nd ed.
ISBN 1-896300-53-7

1. Human ecology. 2. Biotic communities. I. Title.
GF512.W4R68 2002 304.2 C2002-903899-5

Editor for the Press: Don Kerr
Cover Art: Harry Savage
Interior Design: Erin Creasey
Author Photograph: Trevor Harrop

NeWest Press acknowledges the support of the Canada Council for the Arts, the Alberta Foundation for the Arts, and the Edmonton Arts Council for our publishing program. We also acknowledge the financial support of the Government of Canada through the Book Publishing Industry Development Program (BPIDP) for our publishing activities.

NeWest Press
201-8540-109 Street
Edmonton, Alberta
T6G 1E6
(780) 432-9427
www.newestpress.com

Canadian Parks and Wilderness Society
Henderson Book Series No. 12

Henderson Book Series honours the kind and generous support of Mrs. Arthur T. Henderson which made this series possible. The Canadian Parks and Wilderness Society gratefully acknowledges Mrs. Henderson's support of the Society's efforts to promote public awareness of the value of Canada's park and wilderness areas.

1 2 3 4 5 06 05 04 03 02

PRINTED AND BOUND IN CANADA
THIS BOOK IS PRINTED ON ANCIENT FOREST-FRIENDLY PAPER

to Julia

TABLE OF CONTENTS

Agriculture

Trade And Travel

Envoy

AUTHOR'S NOTE

The essays in *Home Place* centre on the theme of Earth as the life-giving, life-sustaining milieu of people. This outward-looking viewpoint is the corrective for in-turned concerns about self and society fostered by the Western tradition. Nature's beauties and bounties have made possible humanity's ideas, social institutions, artistic achievements, and scientific knowledge. Yet the Planet's improbable air-water-land matrix—enveloping home to myriad organisms as well as to humans—has been taken for granted and treated with disdain. "Attend to context—or else" is ecology's primary warning.

In this new edition a few corrections and up-datings have been made. With chagrin I noticed that in the original text taken-for-granted "earth" was often written without a capital E. Strange that the planets—Venus, Mars, Mercury, and the rest—are always capitalized while "earth" is frequently given the lower case spelling, presumably from mundane association with soil and dirt, as if the Planet most important to us were of little concern. Also worthy of capitalization are Earth's synonyms: Nature and Ecosphere. The latter term should be interpreted as the whole Earth and not just its organism-containing skin (the biosphere) because the Ecosphere's liveliness is as much due to internal processes—heat, movement of crustal plates, earthquakes, and vulcanism—as to its surface phenomena.

Population numbers have been updated. Since the 1980s when these essays were written, more than 1 billion people have been added to the global population. The pressure of sheer numbers of consuming humans continues to rise, a fact used to justify continued exploitive economic growth. The God of the Market pays little heed to Earth and its non-human creatures.

After a decade of reflection, I believe that the book's emphasis on Earth's other-than-human components is well-placed. But answers to the question of how to make people acutely aware of their ecological status, how to turn comprehension of humanity's place in the Earthly scheme of things outward, remain elusive. Perhaps most promising is the idea that the locus of life lies in Earth and its geographic ecosystems which constitute the source of all living values.

Stan Rowe
24 July 2002

ACKNOWLEDGEMENTS

Many of the chapters are revised versions of articles or talks. They are listed below, citing the occasion on which they were first presented, in the order in which they appear in the book. The titles of the chapters have in most cases been changed from the originals.

The Relict Grassland: *NeWest Review* Vol. 15 No. 2 (1989).

The First 100 Years: *NeWest Review* Vol. 13 No 3 (1987).

Wilderness as Home Place: *Park News* Vol. 23 No. 3 (1987).

The Quintessential Westerner: *NeWest Review* Vol. 12 No. 5 (1987).

Ecology and Popular Science: Chapter 17 in *Endangered Spaces*, ed. Monte Hummel (Toronto: Key Porter, 1989).

Changing the Global Vision: Chapter 16 in *Planet Under Stress*, ed. Constance Mungall and Digby J. McLaren (Toronto: Oxford University Press, 1990)

Nature, Self, and Art: *The Structurist* No. 23/24 (1984).

Growing Up in Granum: *NeWest Review* Vol. 14 No. 3 (1989).

Beauty and the Botanist: Talk to the joint meeting of the Canadian Botanical Association and the Canadian Society of Plant Physiologists, Regina, June 1982. Published in *The Blue Jay* Vol. 40 No. 3 (June 1982).

The Lake Athabasca Sand Dunes: *The Green & White*, University of Saskatchewan (Summer 1986).

Arks Can't Save Aardvarks: *Canadian Plant Conservation Programme Newsletter* Vol. 3, No. 1. (1988), and *Wildflower* Vol. 3 No. 3 (1989).

Crimes Against the Ecosphere: in *Environmental Ethics* Vol. 2 ed. Raymond Bradley and Stephen Duguid (Burnaby: Simon Fraser University: 1989).

Role of the University: *University Forum* Vol. 2 No. 1 (Saskatoon: Faculty Association, University of Saskatchewan, 1983).

Pro-World Choice: *The Globe and Mail* 16 April 1985.

Goals for Agriculture: Talk to Conference on "Goals for Canadian Agriculture, Prairie Christian Training Centre, Fort Qu'Appelle," (1983). Published in *The Ram's Horn,* (1984).

Transforming Agriculture: Talk at a Forum of the same name, University of Saskatchewan (1988).

Prairie Land and People: Talk at the Spring Festival, the Land Institute, Salina, Kansas (1988). Published in *The Land Report,* No. 33 (1988).

Trading in Water: *Briarpatch* (June 1986).

June Trip to China: *NeWest Review* Vol. 9 No. 1 (1983).

China Revisited: *NeWest Review* Vol. 11 No. 3 (1985).

The editors of the above-listed journals and books were supportive and helpful, and I acknowledge with appreciation their encouragement and assistance. But my greatest debt by far is to Don Kerr who persuaded me to write for the NeWest Review when he brought it to Saskatoon and, more recently, urged me to put together a book of articles. Once the project was underway, he counselled me on its organization and criticized each chapter—unsnarling tangled sentences, identifying lapses of logic and discovering themes of which I was unaware. He was unfailingly generous with his time and good-humoured advice, gentle with his criticism and tolerant of ideas with which he did not fully agree. I thank him profoundly.

FOREWORD

I have a bias. Stan Rowe is my teacher. I never had a formal class from him, but if I had to pick three people whose broad thinking have affected my professional work the most, one of them would be Stan Rowe. Here are two examples of the quality of his thoughts which have informed the lenses through which I see the world.

One is his comparison of the outside view of a unified Earth with the inside "common sense" view that shows all sorts of apparently separate things. The latter is the view of most of us. A good comparison is to imagine ourselves small enough to enter a cell and look around with binoculars. There we would see things like crystals as non-living and things on the move as living, but insight from the outside shows that all are essential parts of the living.

Just so, all Earth is alive, not just the biosphere, which is a less useful term than the inclusive organic/inorganic Ecosphere. Both biosphere and the attendant idea of biophilia come from an organism-centered view, a faulty view that the inside look sponsors. This bio-bias leads us to regard the inorganic world as what Stan calls "loose stuff lying around that we then tend to play fast and loose with." My world has been different since Stan pointed this out.

Here's another shift he brought. Before the end of "biosphere" and "biophilia" as organizing terms came crashing down for me, Stan had written a paper in which he developed the "volumetric criterion for thinghood." This insight he drew after reading J.H. Woodger, a disciple of Whitehead's "organic" philosophy, and James K. Feibleman on the Laws of Integrative Levels. Feibleman outlined twelve general laws applying to the hierarchy of structure from atoms to molecules to cells to tissues to organs to organisms. What comes after organisms? Scientists proposed a variety of categories. Stan settled on ecosystem. He asked what the other levels in the hierarchy had in common, and noted that it was contiguous volume. Species don't have contiguous volume, and neither do populations and communities. An ecosystem's slab of space/time does have contiguous volume, and so functional concepts as well as Feibleman's laws apply. It follows that the two prime living entities for scientific study are organisms and ecosystems.

These are but two examples of a mind at work. In the pages that follow you will relish the products of this truly original mind where his thoughts are not mere abstractions. They sponsor particularities.

At the Land Institute we are at work to build an agriculture based on the way a natural ecosystem works. We use the native prairie as our analogy. We are perennializing major crops to be placed in mixtures that mimic the vegetative structure of that old prairie. Stan's insights are embedded in every pollination, in every soil sample, in every inventory of the mycorrhizal fungi, in every political discussion. I need say no more. Here he is. Read this book.

Wes Jackson
Salina, Kansas

PREFACE

This is a collection of essays on human ecology, comprising a selection of articles written over the last decade. All of them, in one way or another, circle around the concept of Earth-as-Home-Place, an idea whose crystallization began from a chance encounter with a book back in 1947.

Just when World War II was heating up, I graduated from the University of Alberta. My conscience and home influences—a gentle mother and a passionately anti-war father—led me to oppose the draft and, as a conscientious objector, I served in various Alberta and British Columbia facilities—some more confined than others—that gave me plenty of time to think about an errant past, the precarious present and possible futures. At the end of the war, encouraged by my mentor Dr. E.H. Moss, I opted for a Master's degree in plant ecology at the University of Nebraska.

Browsing one day in the library, I came upon an enlightening book by a man named Woodger. Perhaps it was pure serendipity, but I prefer Doris Lessing's idea that few important things happen to us by chance. We are surrounded by limitless possibilities waiting to reveal themselves whenever we are prepared and receptive. And so (I like to think) at just that time, I was ready to soak up, via Woodger, a bit of Whiteheadian organic philosophy and it set me on the track of what became my first and most substantial article in a scientific journal. Titled "The Level-of-integration Concept and Ecology," the paper was written in 1960 and published in 1961, fourteen years after the seed was sown. Almost everything that I have written since then is an expansion and variation on the 1960 theme, ringing its many changes.

In parentheses, I came by my talent for repetition honestly. My father, a Methodist/United Church minister who proclaimed the Social Gospel with prophetic zeal and vigour, used to say he had only one sermon that he dressed up in different words and preached every

Sunday of his professional life. He also told me, with a twinkle in his eye, of sympathy for a brother preacher whose sermon he claimed to have found left behind on the pulpit. Written in red ink in the margin of one paragraph was the reminder, "Argument weak here, shout like hell!" I choose to believe that repetition is not a form of shouting.

The idea I chanced upon is simple enough. The universe and world can be understood as a hierarchy of organizations from simple and small to complex and large, like Chinese boxes or Russian dolls that fit within one another. Atoms are at a low level of organization, while molecules—composed of atoms—are a higher level. Above complex organic molecules are cells, tissues, organs and organisms, a biological series of increasing integration and complexity. If we ask what organized reality exists above organisms, logic points to Earth-space as the last and largest terrestrial "Chinese box" that contains, holds together and supports all the others. People, other organisms, Earth's air, soil and water, are parts of this greater unity and their roles, niches and purposes must in fundamental ways be related to it.

The word we now use for the unit whole that contains organisms is "ecosystem" and it was coined in 1935 by Sir Arthur Tansley, a British ecologist. He had anticipated the importance of Earth-space as a supra-organismic system, and he urged ecologists to attend to this "fundamental unit of nature." Of course! For it is the substantial thing that stands behind and makes sense of the blurry term "environment."

The idea of three-dimensional ecosystems as real objects in Earth-space marks a critical change in language and in concept. "Ecosystem"—meaning home-system—is a physical place surrounding us, to which we belong. In contrast, "environment," as that which is centered on ourselves, has encouraged attempts to pull it into the circle of our belongings, as *our Heritage*. In reality, we belong to the encompassing world and sooner or later it claims us.

In 1948 I joined the Canadian Forestry Service and heard of Angus Hills who had hit on the idea of studying forests as "total sites," his term for ecosystems. A soil surveyor in Ontario's clay belt, he had noticed that tree growth responded not only to the soils in which their bottoms were rooted but equally to the aerial climate in which their tops were "rooted." They were growing within two-layered ecosystems. Landform—the surface shape and kind of geological material—controls both soils and local climate, providing an integrating means of describing and classifying "total sites." Here was a practical application of Tansley's idea, and we began to study and map forestland ecosystems according to landform and vegetation cover using aerial photographs. And so through research, forestry reports, conferences and land inventories, the ecosystem concept came alive and its implications for seeing the world in a new way became clearer.

Study of ecosystems, ourselves inside them, is more difficult than studying organisms. Nevertheless a new awareness of the planet is encouraging science to focus on the Ecosphere and on the land and ocean ecosystems that it comprises. James Lovelock's Gaia hypothesis, which conceives the whole Earth as a vital unit, is popularizing the idea that life is far more than individual organisms. After centuries of neglect and mistreatment, the Global Ecosystem is coming into its own, revealed as the highest-value object we can directly know. The Ecosphere is beyond people, larger than life—as judged by precedence in time, inclusiveness, complexity in organization and evolutionary creativity and diversity.

Home Place is Earth Space. Such is the subject of the book. It is about people in the world and the world around people, about ourselves, astonishingly inside a marvellous Being, enveloped by the Ecosphere. Not only are we in the Earth-envelope, we are parts of it, participants in it, born from it, sustained and reproduced by it. To really grasp this symbiosis is to change radically our appreciation of humanity in the world. In our mind's eye we will see the Ecosphere, see ourselves as willing constituents of it, appreciating the creative bonds that join us to it. And what we know—not superficially but in our hearts and imaginations—has great power over how we act. It is the basis of "political will for change" whose lack today reflects the absence of inspiring visions and elevated goals.

Eons ago all of us—then only potentialities in our ancestors—

lived in the water. Later we moved into the air. The shift made us no less dependent on the planet's skin. In succeeding ages, evolving as humans, we were wanderers in the wild, close to the elements, until we learned to tame animals and plants, to build walls and live indoors. Then our dependence on the Earth seemed less, though really it was as great as ever. The built environment was borrowed from Earth, and our needs for the support of sunshine, water, air and other creatures continued undiminished.

As knowledge grew, we were able to make our lives easier: warmth and shelter more assured, travel less arduous, food supply more dependable. We found that land and water under management could provide for larger and larger numbers of our kind, and so all fertile forestlands, grasslands and waters were turned to that primary purpose. When wealth and leisure were secured, we indulged our artistic and creative talents, surrounding ourselves with useful and pleasurable manufactured things. The wild world receded before a swelling tide of humans and their artifacts, its importance less and less evident—except as a provisioner of raw materials and resources.

In cities, the intuitive sense of ourselves as related ecologically to the land slipped away. The great thinkers of the race—theologians, philosophers, scientists—expressed our isolation from the rest of creation not as loss but as reason for pride. We were "a little lower than the angels," "made in God's image," "innovative geniuses," the elect, a special creation. We named ourselves Man the wise, *Homo sapiens.*

Blinded by our own importance, we were slow to diagnose ourselves, our numbers, our destructive activities, as the cause of environmental problems, which in the first decade of the twenty-first century are building to a crescendo. Still, the majority view is of Earth as separate from us, our oyster, grist for our economic mills. Material economic growth continues to be the watchword, although now the profits must be painted green. The directions of science, the tool of our dominance, are largely unchallenged. Hindsight that tells in glowing terms how far we have come, and how magnificently, bedazzles our foresight.

We *are* a unique kind of animal, conscious, able to reflect on what we do, gifted in wondrous ways. But at the same time, we are tied tightly to the surrounding ecological system from which our talents, physical and psychological, are drawn. Ignorance of our source and, too often, disdain for it, lie at the root of humanity's major predicament. Politics and economics continue to centre on the individual and the collectivity, on free enterprise and social welfare, neglecting ecological necessities of a higher order. Neither philosophical liberalism champ-ioning liberty nor philosophical socialism championing equality will save us from ourselves. Human history will end in ecology, or nothing.

The old astronomers, looking up to the night sky, conceived the world as the centre of an array of crystal spheres moving musically and harmoniously as they carried the stars in their seasonal courses. Today we think we know how galaxies and stars and the planets in their orbits move and are harmonized. We have yet to learn to make music with the one celestial body that is closest to us, to harmonize with the miraculous sun-circling sphere on which our future rides.

VIRGIN PRAIRIE, FAREWELL

THE RELICT GRASSLAND

Years ago, sending a group of students off to the Northwest Territories to undertake various research projects, I expressed the conviction that, regardless of research results, their time and effort would be amply repaid if they soaked up empathic feelings for the North and came back converted—lovers of the subarctic wilderness. They looked at me strangely, thinking of black flies, tinned food and wet tents, and went sadly on their way, shaking their heads.

Is the faith misplaced that all of us, under the patina of civilization, share an instinctive affection for the natural world, a love struggling to get out and find expression, a love that can save us from ourselves? My hopes are rooted in personal experience. Recently a friend, Nikita Lopoukhine, ecologist with Parks Canada, gave me a copy of *Plants of Riding Mountain National Park.*[1] The book's attractive cover—yellow coneflowers on a leafy green background—triggered a rush of nostalgia, carrying me back to the fascinating island of boreal forest that spills over the edge of the Cretaceous escarpment in western Manitoba—Riding Mountain National Park—mine to explore in the summer of 1948 and the site of an unexpected revelation.

All through my young life I had a recurrent dream. In sleep I came into a lovely land, somewhere off to the east in a soft light, with rounded hills and waving grass and the smell of wild flowers in the wind. Each time the fleeting dream was fresh, and its special elation and enchantment was the certain knowledge that none had been there before me. I was the first migrant come wandering out of the west, and all secrets of this primeval prairie, all its mysteries and beauties, lay unspoiled before me.

Born and brought up in the grassland region of southern Alberta, I could never trace this vision to any particular source. Perhaps it came from early beguiling intimacies with the Porcupine Hills where sometimes we went for a picnic in the spring when the slopes were purple with crocuses. Perhaps from the grassy banks of old melt-water coulees and stream valleys, Willow Creek and the Oldman River that we explored as children. Perhaps from a source more ancient, resonating with the spirit of those who crossed the Bering Bridge, came down the ice-free corridor along the foothills of the Rockies and gazed eastward with a wild surmise at the vast unpeopled plain, an Eden of gentle giants: mammoths, ground sloths, bison.

During student days, travelling through the prairies, the dream was stirred by encounters with unfamiliar highlands whose names breathed mystery: Hand Hills, Sweet Grass Hills, Wood Mountain, Cypress Hills, Old-Man-On-His-Back Ridge. Sometimes its memory was quickened by shadowed moraine uplands under drifting cumulus clouds or by distant hills lit by a low sun—scenes rendered more poignant in the golden light by companion Scotty Campbell's favourite song at the end of the day: "All things come home at eventide, like birds that weary of their roaming. . . ."

But evening's seductive promises usually proved empty in the morning. Arrival at distant places, where sunset's glow and long shadows had raised expectations, revealed truck tracks, telephone poles, barbed wire fences, overgrazed pastures and bawling herds of shit-smeared herefords. The few localities that came close to fitting my dream—like the flowery meadows on the top of the Cypress Hills—had already been preempted by others and showed too many signs of use.

Back at the University of Alberta my interest in the prairies was fostered by the enthusiasm and friendly support of Dr. Ezra Moss who provided both recommendations for summer employment that sent me out from Swift Current on community pasture surveys and, following the war, encouragement to take post-graduate studies of grassland ecology at the University of Nebraska. After two years there, Master-of-Science sheepskin in hand and eager for a job, I canvassed all range research stations in western Canada without success. None wanted a green grassland ecologist.

Fortunately the Canadian Forestry Service was looking for ecologists to study the big plants we feed to lumber and pulp mills rather than the little ones we feed to cattle. And so in May of 1948 I came via Winnipeg to the Forestry Research Station on the Riding Mountain, metamorphosed overnight from grassland ecologist to forest ecologist. Goodbye grasslands, thought I. The judgement was premature.

Ten thousand years ago, when the continental glaciers were melting back northward in a warm dry climate, prairie vegetation dominated the belt that today is occupied by the aspen, birch, spruce, pine and larch of the southern boreal forest. Five thousand years later the climate cooled, rainfall increased, and forest invaded the grasslands on all except such warm dry sites as sandy and gravelly landforms and south-facing slopes. In these sunny sanctuaries the native herbs and grasses survived.

As I later discovered, many tracts of the old grassland lie scattered through the western half of the Riding Mountain, the larger ones—even though in a National Park—used and abused as grazing lands by surrounding settlers. But in out-of-the-way places some survived in their natural state, protected by wetlands or far from trails, even within the higher, more heavily timbered eastern part of the Park. And one June day, not even suspecting that such things existed within the forest matrix, I stumbled on a jewel.

My assignment was to keep track of a boisterous student crew who, between such death-defying acts as chasing black bears, hassling hornets and peering into goshawk nests, were employed in marking out a sampling grid in the spruce-aspen forest just north of Wasagaming Lake. One-tenth-acre plots were staked on east-west lines across each section, and so we advanced, surveying new territory every day.

That memorable morning, reconnoitering ahead of the crew, I sensed that the shadowy spruce canopy overhead was lightening down the line, and vaguely the parallel illumination dawned that I was approaching a forest edge. Could it be a wetland, a muskeg? No, the terrain was too high. Something strange lay ahead, under the sunshine and blue sky.

I plunged through the trees, pushed aside the fringing bushes and, heart pounding with excitement, entered as close a match to dreams as reality can ever afford: a wonderful prairie island in the forest, little hills and valleys bright with wild flowers, grasses waving in the breeze, sweet meadow smells.

And mixed with the elation of discovery and all the first joyful impressions, a strong sense of affection, of coming home to my grassland-in-the-trees and belonging there.

I remembered another peak experience when, physically tired, I came into a warm room of friends and relaxed into a sudden radiance. Such moments, when the golden bubble bursts, signify health in its meaning of wholeness. For a brief spell we are harmonized not only with what surrounds us but also within—mind and emotion, thoughts and feelings. That we can experience "highs" with good friends and respond similarly to the beauty of wild nature lends hope for a more sensitive and appreciative society in a world more sensitively and appreciatively known. The cooperative community and society has in fact been approximated at various times and places. Now to get in tune for the next step, the cooperative ecosystem, setting aside species selfishness and our utilitarian bent, with heartfelt charity for the Home Place.

❧

Sometime soon we will recognize the difference between going to a native grassland to sit and listen and learn from a microcosm of the World, to open ourselves to it for inspiration as to how best to live with it and minister to it, and going there with the intention of turning it into a source of recreation, into a show place for wild animals, into a landscape painting, into a pasture for cattle or—most terrible thought—into just another wheat field.

THE FIRST 100 YEARS: LAND USE IN THE PRAIRIES

The ways that we use and exploit the Earth's surface are direct and visible measures of sensitivity to our source. The landscapes of our making match and reflect society's cultural inscapes. Land use, mirroring our North American minds, cries out for changes in attitude toward what surrounds us.

Little more than a century ago the shift in land use from hunting-and-gathering to agriculture was initiated on the Western Plains. Today the prairie scene is completely changed from what it was when Canada's premier naturalist John Macoun came botanizing by, exclaiming at the marvellous flowers while also declaring the grass lands fertile and suitable for farming. As we look out from the rectangular lots and fields that enclose us today in town and country—the legacy of the grid land surveys of the 1870s and 1880s—we find it difficult to imagine the curvilinear sights, sounds and smells of the primeval grasslands, now reduced to a few forlorn and untypical fragments.

ATTITUDES SHAPE USES OF LAND

The land-use changes that began toward the end of the 19th century were no accident. They were the expression of European attitudes and perceptions of the prairies—occupied sparsely then by hunters-and-gatherers—as *nothing but* wilderness, waste, barren, desert and deserted until colonized and "improved."

The rapidity of the land-use transition reflected a Canadian National Policy promulgated in 1879 in response to several political urgencies. One was the perceived threat of an American expansion, to be countered by the establishment of a communication and transportation system that would make possible—through settlement—the occupation of the Prairie Region. Another was the ambition to create a new and relatively exclusive market for the expanding financial and industrial productions of Eastern Canada.

Railways, tariffs and immigration were parts of the same package, aimed at contributing both to the military protection and to the economic integration of the new Dominion. And times were opportune to sell cereals abroad, particularly the seeds of the desert grass, wheat, because the booming Industrial Revolution in Europe continued to boost the growth of urban labour forces and their demands for bread.

Prairie settlement was born into a capitalistic era of machine agriculture. It was a new start, marking a break with older village-centred feudal forms such as existed in Europe and Asia. There the constant threat of raids and of warfare and the lack of police protection and dependable water sources had encouraged rural people to cluster together in villages from which they travelled out daily or seasonally to engage in subsistence agriculture.[1]

The comparative safety of Western Canada as well as the methods of allocating land to the railways, land corporations and individual settlers had the effect of scattering farmsteads, with towns established at grain-shipping points. The latter, usually ten to fifteen miles apart on the railway lines, were spaced to suit the capabilities of horse-powered transportation.

From the beginning, farming was oriented mostly to the market, to raising food for export, for cash. Ideas of traditional subsistence farming, and of regional self-sufficiency in food, were left behind in the East.

The early insinuation into Prairie agriculture of the idea of farming as a *commercial enterprise*, as a business rather than as a provisioner of food for domestic consumption, lies at the root of the exploitive land uses that continue to plague the West today. For attitudes as to what farming means—the cultural goals of agriculture—have much to do with the way rural society is organized and how the land is managed and conserved, or mismanaged and degraded. Attitudes have determined land use on the Prairies for a century and only profound changes in attitudes can change, for the better, the land uses of the future.

Despite rhetoric about farming as primarily a way of life, and only *after that* a money-making enterprise, the family farm in its old independent sense is a nostalgic relic, a museum piece. The

Jeffersonian farmer and husbandman, self-sufficient and secure on his small holding, the bulwark of democracy because of his rugged *unsubsidized* independence, is long gone.[2] Criticized in Regina in 1987 for supporting the protectionist policies of the European Economic Community, France's President Mitterand countered with the assertion that, on the average, each Canadian farmer is subsidized by about $30,000 yearly.

ECONOMIC GOALS SUBVERT CONSERVATION

Massive industrialized production on ever-larger farms is the present goal, and development economists set the standards to which most farmers adhere, willingly or not. According to current theory, progress is gauged by productivity figures, that is, by output per person. Applied to agriculture as just another business, progress is therefore inversely related to the percentage of the population engaged directly in agriculture. The fewer the farmers on the land, the more advanced the economy, on the assumption that reduced employment in primary production—in farming just as in forestry, fisheries and mining—frees up labour and capital for more worthy industrial and service pursuits in the cities.

The land-use changes that have accompanied the intensified industrialization of Prairie agriculture since its beginning are well known. The average size of farm has steadily increased as the numbers of rural inhabitants and the areas of natural landscapes have decreased. Energy-intensive technology, in the form of the tractor, made horses obsolete on the land, and, in a similar way, powerful machines for tilling, seeding, spraying and harvesting have also made people obsolete on the land.

We sold the horses for meat but we have not yet figured out what to do with excess people. Agriculture has led the way in jobless growth, and high levels of unemployment that used to be a social aberration are now accepted as a normal part of the socio-economic system. The more ingenious we become at abolishing work, the more the promise of job-creation is the focus of politics.

In the West, mixed farming or diversified farming has not been successful "because of the loss of efficiency compared to specialized production," to quote a still timely Report on Manitoba's Economic Future.[3] In the attempt to coax increased production from the soil and so beat the cost/price squeeze, the landscape is more and more simplified in bigger and bigger chunks.

Largely unrecognized is the fact that the cost/price squeeze is the heart and soul of the industrial agriculture system, both driving it and setting its priorities. Farmers' costs are the profits of agri-business, so naturally the pressure on them is always up. The prices paid for farm produce must be competitive in the world market, so the pressure on them is always down. The cost/price squeeze is here to stay. Yet farmers are encouraged to believe that it will disappear in some happy future and that meanwhile it can be beat: by more efficient farming; by computer-guided management; by higher yields; by doctoring and force-feeding more land, by intensifying the industrial business approach.

Industrialization mania is a sure formula for the destruction of rural communities as well as for the draining of wetlands, the clearing of woodlands, the abuse of marginal lands, the deterioration of soils. But convention continues to intone, "Have faith. Don't doubt. Bigger is better. Industrialized agriculture on its present course is Progress."

UNCULTIVATED LANDS THREATENED

Is this the full story of land use over the last 100 years? Is it fair to imply that we are simply a non-conserving people? What about all the *uncultivated* land of the Prairies—the rangelands and the aspen parklands—offering the promise of a balance in uses alternative to till agriculture? Or is it, too, ticketed for "improvement" by the plough?

According to Dr. R.T. Coupland, the belief is erroneous that land not already in arable agricultural use will remain so, with such factors as stoniness, rough topography, salinity, sandiness and droughtiness guaranteeing immunity from drastic disturbance. On the contrary, the Canada Land Inventory figures for Saskatchewan indicate that irremediable impediments to cultivation exist on only *5 per cent* of the land surface in the grassland zone. In theory about *95 per cent* of southern Saskatchewan could be tilled (including, for example, the site of the Grasslands National Park) either for cereal and oil-seed crops (approximately 60 per cent of the land surface) or for "improved pasture" (approximately 35 per cent of the land surface). The area on which these calculations are based includes Indian Reserves and Crown Lands, but excludes provincial parks, major water bodies and urban areas (which are small in total).

Quoting one agronomist, "We in the West have not yet exploited our agricultural potential to the full. There are 35 million acres of Western Canadian land in Canada Land Inventory classes 1 to 4 as yet unused."[5] Most of such potential farmland is in the Peace River country; little good land is unploughed in the southern region.

Not unexpectedly, agricultural policies have been directed toward the exploitation of the remaining untilled areas. Subsidized programs of "improvement," of conversion from the native state, are always popular—during prosperous times to take advantage of favourable markets and during desperate times to provide that extra increment of production or of quota acreage that might make the difference between farm survival and failure.

SHRINKING GRASSLANDS

The unforested part of the prairie provinces occupies only 5 per cent of the land area of Canada but it comprises about three quarters of the land that is tilled every year. Within this western region, something over 60 per cent of the land surface is at present devoted to arable (tilled) agriculture, and, in the normal regime, more than 50 per cent receives some form of cultivation every year. Not only is the degree of annual cultivation high—possibly exceeding that of any other comparable region in the world[6]—but also it continues to expand incrementally. The native prairie rangelands and other wild lands in the southern Mixed Grass zone suffer steady attrition. Little wonder that the West has the dubious distinction of being at the centre of so many "wildlife in jeopardy" programs.

Northward, the aspen parkland with its fescue grassland, is in the same predicament—reduced to a patchwork of small remnant endangered ecosystems in a matrix of tilled land. Exploited and simplified by agricultural use, trends are the same as in the south. Estimates of the natural areas surviving range from a high of 20 per cent to a low of 2 per cent, depending on the subregions considered.[7] The fragments that remain are mostly of small size and are increasingly important as refugia for plants and animals whose surrounding natural systems have been destroyed.

A few major blocks of native prairie are protected in national parks and military reserves. The larger provincial parks, however, with few exceptions, allow multiple uses such as grazing, timber extraction, gravel mining, oil and gas exploration and exploitation and, of course, various kinds and intensities of recreational use. The cumulative effect is to deteriorate and decimate the grassland types. Other tracts of crown land are also exposed to multiple uses with little regard for preservation. Would we treat them better if we called them what they are, "public land" not "crown land"? Community pastures that as "crown" lands should be managed in the public interest—as a bare minimum by protecting their native biota—notoriously reflect a lack of care.

Eastward, Manitoba's Tall Grass Prairie is more than ninety-nine per cent gone, and not until the late 1980s was an inventory mounted to locate the last few acres of what was once a rich and beautiful type. Incongruously, the largest protected fragment—the Living Prairie Museum of twelve hectares—carries a city address: 2795 Ness Avenue, Winnipeg. Under our feet a miniature "Amazonian forest," of which we knew little, has disappeared.

The difficulties in obtaining a balance between agricultural and other legitimate land uses are pointed up by the almost insurmountable obstacles encountered in trying to preserve native landscapes for their own intrinsic values. Attempts, largely unsuccessful, have been made since 1941 in Saskatchewan to protect small areas of grassland permanently for scientific purposes. Negotiations with provincial authorities in the 1960s provided legislation that could have given permanent protection to a number of designated grasslands, but it was repealed before put to use. Again, in the '70s, the International Biological Program generated pressure for preservation of natural areas, and one hundred representative or unique landscape tracts were proposed to the provincial government. A few important tracts such as the three-square-mile Matador Grassland gained the protection of the Parks Act, but, in ten years under the Ecological Reserves Act, the province managed to designate only one small parkland area on the banks of the Assiniboine River.

GRASSLANDS—THE "FAILED RESOURCE"

The problem of attaching importance to natural grassland is partly one of perception. Neil Evernden tells the story of a trip by rail across the Prairies, looking out from the dome car, while the conductor commented, "Don't know why—there's nothing to see."[8] Evernden ironically calls the prairie "a failed resource" because to eyes attuned to forests and lakes, to mountains and sea, it baffles with distant vistas, spaciousness and apparent emptiness, with its big sky, sweep of wind and brilliant sun that threatens, he says, "to bleach the ego."

Many have asked in perplexity, "How can there be an interesting Grassland National Park?" Henry Kelsey, one of the first European travellers on the Yellowhead Route 300 years ago, heading westward toward the Battlefords area like so many tourists since, expressed the sentiment in forgettable doggerel:

> This plain affords nothing but Beast and Grass,
> And over it in three days time we past.

Like Kelsey, the settlers of Manitoba and the Northwest Territories (from which Saskatchewan and Alberta were carved in 1905) came mostly from the milder forest climates of Europe and eastern North America. Grasslands to them seemed to lack an essential ingredient—trees. Perhaps a forestation program could contribute to a friendlier environment? The thought was expressed in early reports that recommended tree planting to embellish the plains.

In his Report of the Forestry Commissioner to the Minister of the Department of the Interior in 1888, J.H. Morgan conjectured that the rich soils of the west were treeless because of their heavy rank growth that encouraged fires. Stop fires and the forests will return. He drew attention to John Macoun's 1880 report of a "desert" in southwest Saskatchewan (the Great Sandhills) where grew large cottonwoods and "a perfect oasis of nearly seven hundred acres surrounded by sandhills that kept out fire." Morgan recommended establishment of Experimental Forest Stations to ascertain the tree species most suitable and valuable, following which "our next duty would be the reserving of large tracts of land for permanent forests." The result of these encouraging words was the tree nursery at Indian Head that began providing planting stock in 1901, and numerous plantations on sandy Forest Reserves such as Shilo, Elbow and Dundurn beginning in 1916.

This culturally motivated attempt at an alternative land use was unsuccessful. The plantations were ravaged by drought, fire and rabbits that showed no respect for the theories of Macoun and Morgan. The grasslands asserted that they were not forestlands. In the general transfer of resources to the provinces in 1930 the federal government surrendered its sandhill forest reserves, doubtless with a secret sigh of relief. Now these sensitive dune lands, disappointing as forest producers, have been made into provincial parks, recreation areas and community pastures. Some have been ploughed and wind-drifted, and a few large blocks—as military reserves—test the treads of tanks, the shock of mortar fire.

UNCHALLENGED TECHNOLOGY LEADS LAND USE

Technology continually directs the way land is used for crops, for transportation, for whatever the Earth's crust can be made to yield. Applied to agriculture, its goals of efficiency and risk-reduction move farm practices from hoe-and-sickle to horsedrawn plough and reaper, then on to huge air-seeders, monster combines, massive doses of chemical fertilizers, dangerous biocides and all the other techniques that go with and justify large monocultural fields.

In a parallel way, transportation technology—another important determinant of land use—has evolved over the century from travel by horses and railroads to the automobile, airplane, helicopter, all-terrain vehicle. Again, as with agriculture, the face of the Earth is modified more and more drastically, and the changes are not friendly to native landscapes.

The technology of mining that provides energy and materials for industrial agriculture has followed a similar course. First the near-surface materials were exploited: gravel, salt, sodium sulphate and lignite. Then came underground mining of bituminous coal, potash, metals, and deep drilling for gas and petroleum. Each technological "advance" has required greater expenditures of fossil fuel and hydro-electric energy, and each is marked by greater environmental damage.

The ever-changing imprint of the technologies of agriculture, transportation and mining on western landscapes reflects government policies and individual decisions that follow a taken-for-granted, unchallenged agenda: to make life increasingly better, easier and more comfortable for people. From a single-species standpoint, this is a thoroughly commendable goal toward whose achievement technologic expertise is marshalled. From a broader perspective, the idea is terribly flawed. A single species, even a smart one, cannot go it alone. We cannot continue to appropriate *everything* for ourselves because the land that we use is also the ecosystem in which we live, the life-support of ourselves and other organisms.

Efficiency and risk reduction in the service of humanity, the inner motivations of technologic change, must be rethought from an ecosystem perspective.

MINING: TECHNOLOGY GONE ASTRAY

The example of mining as a land use is particularly instructive for it shows clearly the dangers of technology unguided by ecological wisdom. Delving in Mother Earth has always had its critics, those distressed at what seems desecration, pained at the mining of her veins. Such discomfort, we are finding, is completely justified.

Terrestrial organisms evolved in the world of renewable and replenishable resources, in the post-Precambrian green world of simple plants, clean water, organic soil, fresh air. The hostile environment that had existed earlier in the pre-organic period of Earth's geological history was succeeded by a safer environment, screened by ozone from excessive exposure to ionizing ultraviolet rays from above, blanketed by sediments from excessive exposure to the hot radioactive rocks below.

As time went on and the Earth evolved, dangerous toxic substances—heavy metals, sulfurous compounds, radionuclides and hydrocarbons—were safely sequestered beneath the planet's surface. These we call "non-renewable resources" and dread the day when humanity will be without them. We should call them "Unnatural Resources," mark them with a skull and cross-bones, and pray for the day when humanity will choose to do without them. Their use is at the root of most serious pollution problems, actual and potential.

During the last few centuries, through mine shafts and bore holes, the technology of mining has introduced the Unnatural Resources in massive amounts back into the Earth's life space, depleting the ozone layer and poisoning air, water and soil. In a sense, mining and the modern use of mining products are turning the geological clock back, recreating the Precambrian environment when acid rain, laced with radionuclides and heavy metals, washed the rocks of a world hostile to life. Yet today, like idiots, we stand and stare at accumulating toxic wastes, at the dying fish and dying forests—clear signals, like the belly-up canary in the mines haft—and ask: How can this be? Surely the technology of non-renewable resource use is good?

The short answer is *Rubbish*, though exceptions can perhaps be made for a few mineral forms such as phosphorus and potash, two of the life-enhancing substances dug out from underground. Even these deserve careful attention in exploitation and use. Phosphorus should be recycled because it is relatively rare in usable forms. Instead, we routinely flush our phosphorus-rich wastes into the rivers. As to potash, we have no idea of how to handle, in perpetuity, the salt heaps—two tonnes for every tonne of fertilizer—that are accumulating massively beside the mine pits in southern Saskatchewan and elsewhere.

Technology seems to have taken on a life of its own, forcing its mechanical, non-organic efficiencies on the landscape, destroying in the process the world's priceless variety and wildness. Yet technology comprises human inventions, susceptible to guidance once goals are clearly defined. In the case of mining, for example, a better goal would be the extraction and use of those surface Earth minerals and substances to which evolving life has long been exposed. An industrial world based on safe ceramics is preferable to the dangerous one we seem bent on reinventing.

Similarly with agriculture, is its primary goal of maximum production reasonable and life-enhancing? In a world awash with cereal grains and vegetable oils, will Canada continue to add to the surpluses whenever the climate turns favourable, depreciating what remains of the fertility and beauty of Prairie landscapes, undermining with cheap grain the indigenous agriculture of less wealthy countries?

CONCEPTS AND ATTITUDES TO GUIDE LAND USE

Problems with land use are, to a large extent, the reflection of failed values, attitudes, concepts, which the inertia of technology perpetuates. One hundred years of land use on the single track of high agricultural productivity will not be side-switched, let alone reversed, without radical changes in the underlying motive forces.

Two related ideas could bring a more balanced perspective to land use, in Canada and everywhere in the world. One is a concept, the other an attitude, and the land is hurting because of their absence. The missing *concept* is the ecological one of *landscapes-as-ecosystems,* literally "home systems," within which organisms, including people, exist. We have been taught that we are separate living *things,* surrounded by other living *things,* but not so. The realities of the world are ecological systems of which organisms are components and without which no creatures of any kind could exist. The biggest ecological system, the planet or Ecosphere, is composed of regional and local landscape and waterscape ecosystems of which life is one property. Living on the land, under the sky, we people are inside the prairie landscapes, inside the continental ecosystem, inside the Ecosphere. The health of each and all is our health.

Conservationists and preservationists spend too much effort on threatened species, as if (next to us of course) they are the only important things on God's Green Earth. True, the importance of preserving wildlife habitats is more and more recognized as essential to the species which cannot exist without them, but "habitat," like "environment," is a weak and woolly concept. Both fail to define the real three-dimensional ecological systems in which we are immersed; they fail to project to the public the sense of importance of the enveloping natural world and the urgency of protecting it.

The task is to think of ourselves as within ecosystems, from the big one shown in satellite pictures—the Ecosphere, the Homesphere—down to the small landscapes that it comprises, those that regionally and locally support all existence. These enfolding land-and-water systems, used and abused by humanity, are more than resources; they are part of the miraculous world ecosystem that brought life into being, sustains it and renews it.

The missing *attitude* is sympathy with and care for the land and water ecosystems that support life. It will come when we make the concept of a planetary home part of our daily thought, part of our hearts and imaginations, after we have conceived of landscapes as real three-dimensional things to be valued and have recognized their importance throughout the educational system. Human beings, incorrigibly species-centred, have difficulty conceiving that things other than themselves (with the exception of some look-alike animals) merit compassionate attention.

In the big-sky west where the light is bright, the distant vista exposes patterns of land uses, similar over hundreds of miles. Are we doing right by the land? Few question current trends. Yet here, if anywhere, humans should clearly see their roots in the land, understanding that they are *from* the land and belong to it in a way that it can never belong to them.

The opening of the twenty-first century, a new millenium, is an inspiration to change our ways in western Canada. Even now, soils are drifting, valley bottoms are being broken, wetlands are still being drained, aspen bluffs levelled, patches of native prairie ploughed. Hopeful signs are the many recent conservation and preservation programs in response to popular demand, supported by both non-government and government agencies. After years of inaction, *they all sound so good!*

Let us, however, remember as we support them that without parallel changes in how we think and feel about the land, these and all such conservation/preservation programs will prove to be only stop-gap measures, finger-in-the-dike exercises, glimmerings of hope, but not the new day dawning.

WILDERNESS AS HOME PLACE

Newfoundland's irrepressible John Crosby advanced a thought that merits deliberation. "Ten years ago we didn't know about environment," he said, "but now it's all around us!"

It's true. Each of us is born, lives and is recycled in a marvellous surrounding environment that has been so un-sensed, so taken for granted, that until recently we have been unaware it's all around us. Furthermore, we didn't really care.

Who worries that each fall the willow rings around prairie sloughs are fired, preparing the sites for ploughing if the succeeding year is dry? Who mourns the demise of scattered aspen bluffs smashed down by bulldozers in the winter when the green-wood is frozen, then windrowed for summer burning? Who laments the passing of the wild grasslands, skinned then engrafted with tame wheat and domesticated pasture?

Resistance to respecting the inherent worth of the natural world and protecting its diversity is deeply anchored in Western culture. The famous men we praise—the great theologians and philosophers, the eminent artists and scientists—with few exceptions shared one blind spot, one major intellectual defect. They knew practically nothing about the relationships of the human species to the Ecosphere. They conceived the Earth as resource not as source.

Read the sacred scriptures, study the works of the cultural giants, and an overpowering conclusion emerges. Only two relationships are important: that between Man and God, and that between Man and Man. The Man-Planet relationship is simply not recognized. At the most, Nature is seen as a providential provisioner. At the least, Nature is only a scenic backdrop on the stage where Man, proud Man, plays out his dramas.

Lest the gender-conscious protest that the other sex has been slighted by references only to Man, the masculine term is purposely used because the other blind spot of Western sages is precisely the feminine. They have paid as little attention to Woman as to the Ecosphere. As an extension of what was expected in their homes, they projected an inert Nature, a passive Mother Earth, simply there to be *husbanded*. Today of course even husbandry is passé. We "manage the resource base optimally."

Nothing is to be gained by dwelling on the ignorance of our otherwise illustrious ancestors. That they were not great ecologists is regrettable, but for that they cannot be blamed. As little as four or five centuries ago, Europeans thought that the Earth was the centre of the solar system, and only in the last 50 years have suspicions dawned that things other than people may possess intrinsic values. We, however, *have* been granted glimmerings of ecological insight, enough to know that the answers to current environmental woes will not be found in the pronouncements of the ancients. Theistic or humanistic, but always homocentric, they speak to the human condition as it existed before 1950. They speak for a world that excluded the Real World.

Today we are in an entirely different game, recognizing a new supra-human partner—the Ecosphere, Gaia, the Natural World. We are trying to understand realities more important than people, trying to escape the exploitive macho approach that aims to rape and rip-off the non-humanized world, trying to learn the new symbiotic rules because the old ones are killing us—those that seemed to work so well when we were fewer in number and less powerful, before we equipped ourselves with huge machines and appropriated massive pools of fossil fuels. Now we are struggling to understand what it might mean to become *compliant cooperators* with Earth's ecosystems, hitherto insensitively appropriated as *our* resources and *our* heritage. What monumental conceit! What egregious insolence!

Whether evolutionist or creationist, no one can deny that we are Earthlings, out of Mother Earth and her moon rhythms, our bodies composed of the surface substances that lie ready at hand: star dust, humus, air and water. Notably, we are *not* composed of heavy metals, radionuclides, petroleum hydrocarbons, nor any of those unnatural resources that we persist in digging out from underground to poison and pollute the planet and ourselves. Our make-up suggests that we should avoid such toxins as unnatural.

But what *is* our nature? It seems logical to assume that whatever our essence may be it somehow shares in that greater Nature from which and into which our species first came. We were born a wilderness primate and for eons lived by harvesting the sun on land and in water, foraging among the world's forests, grasslands, streams, lakes and seashores.

So the question, "Why is wilderness important?" is like asking why anything in our background is important. Are calcium and phosphorus of which our bones are made important? Is fresh water important? Are trees and flowers and grasslands important?

Some will answer "yes" and call the questions silly. But many, lacking a lively sense of biological beginnings and ecological history, are unconvinced. Content to accept the increasingly mechanized world in and around cities as sufficient for human needs, they ask: Have we not risen above the animal world and its wilderness trappings by renouncing Nature?

We should in turn ask these people: Can you understand yourselves apart from your ethnic roots, your cultural history, your family ancestors? So also as Earthlings, look to your beginnings, to your *natural history,* for insight to your needs, your interests, your feelings, your sense of beauty and belonging.

Perhaps by saving wilderness we preserve an important part of ourselves; That was Thoreau's belief. In wildness, he said, is the salvation of the world. For him, as for Grey Owl, the spiritual dimension of wilderness experience was foremost—the antidote to humanity's preoccupation with itself.

Often, without realizing why, we recreate ancestral environments, surrounding ourselves with green carpeting and floral wall patterns, caged birds and pet animals, landscape pictures of trees and water, lawn savannas sprinkled with shade trees and shrubbery. We try to escape monotony by creating a tamed diversity, dimly recognizing that asphalt and glass buildings do not a living world create. Further, we recognize the deadening effects of artificially cheap and mean environments.

Such responses, such memories, are powerful arguments for interspersing, within the monocultural wheat belt, major tracts of diverse native grassland and woodlands too, insofar as they have survived in valleys. Unfortunately, most native prairie has been heedlessly destroyed by the cow and the plough. No thorough inventory was taken before the damage was done, and we will never know the extent of the loss.

Two values have been proposed for the remaining wild natural ecosystems of the Earth. The first is the use-value of what is preserved. Both the World Conservation Strategy and the report of the World Commission on Environment and Development (the Brundtland Report) stress the utility of preservation. Save ten or twelve per cent of continental forest and grassland ecosystems, they say, because we may need them in the coming population crunch. Save endangered species because of their potential for food and drugs. In its best light, the thought is ungenerous.

The second value, on a higher plane, is symbolic. The fundamental reason for preserving whatever wildness remains on land and in water is the symbolism of the act, the implicit recognition of values beyond humanity, something other than ourselves that ought not to be destroyed, an expression of wonder and awe before the marvellous world that created us and that, once gone, we cannot recreate.

Note the shift in emphasis, like changing the popular slogan, "Parks are for People" to the more subtle "People in the Service of Parks." Wilderness, wild areas, can only survive as natural ecosystems that are ministered to, served, not just preserved, for only such attitudes can our part set their values above human needs and wants. Otherwise, bereft of the only protection that counts—high valuation—they will be used to death.

How are we doing on the western plains? Where are we with our new appreciation of wild areas and wild creatures? Just about where everyone else is: picking around the edges of a few dominating land uses—agriculture, forestry and mining—that continue to treat disdainfully what is left of the native landscape.

Agriculture rules the south, and ideas of balanced land use—the essence of conservation—have been thrown out the window. Apart from a few sections of native grassland preserved due to efforts initiated under the International Biological Program of the 1970s and a scattering of smaller reserves, the native ecosystems that once occupied the fertile lacustrine and ground moraine plains of the prairie provinces are gone.

Public unease is countered by old arguments that all the land is needed for farm-family survival, for food, for greater production. People must always be first, the world environment a distant second. If twinges of guilt are felt, we salve them with efforts to obtain more parks as recreation areas, with weak Ecological Reserve Acts, with forlorn attempts to save a few rare and threatened species of animals and plants. The primary focus is on habitat conservation—not in the interests of the wildlife but to ensure that sportsmen are not without moving targets.

We propose to stem the development tide by a multitude of uncoordinated finger-in-the-dike programs with utilitarian purposes up-front. In western Canada, we have, under various degrees of protection and subject to the vagaries of ministerial discretion, a range of permitted uses: provincial parks, park reserves, wilderness areas, wildlife management units, fish and wildlife development fund lands, regional parks, provincial wildlife refuges and sanctuaries, game preserves, heritage marshes, protected areas, provincial and Federal community pastures, critical wildlife habitat, public hunting grounds, fur-bearing animal refuges, game bird refuges, goose refuges, ecological reserves, natural areas, buck-for-wildlife projects, prohibited access wildlife areas, special conservation areas, national parks, national historic parks, migratory bird sanctuaries, national wildlife areas, Ducks Unlimited projects and military reserves.

Let no one say that we are not hard at work in hundreds of little ways, throwing up flimsy barricades against encroaching development, trying in makeshift ways to ensure that some Wild survives in the West!

The fortress approach to preserving wilderness will not work. More than seventy threats to wilderness parks have been identified, most originating outside park boundaries: acid rain, mining, air and water pollution from pulp mills, urban encroachment, roads, sour-gas wells, deforestation, overgrazed rangelands, till agriculture, to name a few.

Protective fences are not the answer. The only firm foundation for wilderness preservation is psychological and attitudinal, beginning with the recognition that the ancestral world is *important,* far beyond the importance of its most precocious species.

If current attitudes and trends continue, revisit the Grain Belt in twenty years and experience terminal monotony—no more native prairies, no undrained wetlands, no aspen bluffs, no lakes worth the name, no interesting kame, esker and glacial beach landforms (they are going fast for sand and gravel). Just a few salt-tolerant and drought-resistant cereal and oil-seed crops covering the depleted soils for miles and miles, with rangelands ploughed and seeded to brome and crested wheat grass for miles and miles.

That prospect ought to spur quick action. The stakes are high in terms of landscape attractiveness, good water and food, possibly even survival. Ecocentric education, switching the emphasis away from ourselves, is the key. Parents, teachers, everyone, along with responsible institutions—both governmental and university departments of environment, of natural resources, of parks and natural areas—should turn their attention to it and make it their focus and special mission.

To be at home on the planet and welcome here, humanity must understand and appreciate the primacy of that home, the Eden we have never left, and the wild that is its emblem.

THE QUINTESSENTIAL WESTERNER

Against the background of a quiet northern lake and a darkening sky, Grant MacEwan, silver-haired and in his mid-80s, everyone's ideal grandpa, looks out benevolently with a quizzical eye from the cover of his latest book, *Entrusted to My Care.*[1]

Prairie people, we who exist in the hinterland out on the periphery of Canada, are baffled by the identity crisis of those whose lives have been disoriented through long exposure to the magnetic fields of cities like Toronto. What is this constant talk about not understanding who as Canadians we are? The people that I know have no problem, nor do the westerners that I read about. Does Mel Hurtig have a problem? Does Don Kerr? Does Aritha van Herk? Well, neither does Grant MacEwan.

Here is a good man, at home in the west, who honestly cares about humanity and its future, a man dedicated to making the case for respecting and better managing the environment, one who identifies extravagance and thoughtlessness in exploiting Canada's resources as the road to national decay and ruin.

It can happen here he says, and so "this volume has been prepared with one main hope: that it might awaken a greater concern for the care of resource treasures, of which the Canadian inheritance was great." The word "was" is significant. An earlier version of the book, published in 1966, consisted of forty or so cautionary essays exposing misuse and depletion of resources in the prairie provinces. This latest version enlarges and deepens the original. Thanks to MacEwan's scholarship and historical perspective, his thoughts of the '60s remain topical, providing numerous insights on the attitudes of influential characters whose activities have affected the "development" of the west over the last century. New chapters add pertinent afterthoughts that update conservation efforts. MacEwan concludes his book with a supplication or prayer, expressing the prophetic need for superhuman efforts, if not divine intervention.

The heroes of the chapters are those who worked hard for conservation as they understood it. In 1901, Norman Ross accepted a position with the Department of the Interior to encourage tree-planting on the prairies. He zealously promoted the use of farm shelterbelts and amenity plantings through tree nurseries that he designed at Indian Head arid at Sutherland on the edge of Saskatoon.

Norman Luxton of Banff helped to bring the bison back to Canada, alerting the federal government to the availability of the only sizeable herd of prairie buffalo in the western USA. A deal was completed in 1907 by which "God's unbranded cattle," as the Reverend John McDougall called them, were reintroduced to Elk Island Park and the Wainwright Reserve.

C.S. Noble, father of the poet, combatted soil erosion in southern Alberta in the dry 1930s with his invention the Noble Blade, a cultivator that cuts off weeds underground and leaves on the surface a protective stubble mulch. This was an early example of "minimum till agriculture" that again today is in the save-the-soil news.

George Spence, prairie pioneer and Director of the Prairie Farm Rehabilitation service (PFRA), helped to sponsor 185,000 water-supply projects on farms. He spoke for the majority in opining that "it must be nothing less than a sin to allow needed water to be lost to join the ocean." Conservation in this sense lives on; in 1989 the Premiers of Newfoundland, Quebec and Ontario agreed to cooperate in exploiting hydroelectric power sites in Labrador, stating that it was "an absolute shame to see hydro power running to the sea unharnessed."[2]

Robert Walls, a Calgary schoolteacher, instilled in his students an interest in trees and shrubs. They also benefitted from the discipline associated with Arbor Day planting and with caring for these living things.

In the same vein, MacEwan pays tribute to two men who pointed him toward the; goal of land stewardship: W.R. Motherwell and MacEwan senior. Dr. Motherwell, Saskatchewan's first Minister of Agriculture in 1905, a Laurier anti-conscriptionist during World War I and twice federal Minister of Agriculture under MacKenzie King in the

1920s, "charged me solemnly to use my voice and open for the protection of Canada' s soil, grass, water, trees and other resources."

As for father, says Grant, it was his view that Nature's Gentleman could be identified as the person whose motivating purpose was to leave things better than he found them. "Sharing Albert Schweitzer's conviction that we owe kindness to every living thing, he refused to shoot a coyote and tried to avoid stepping on insects; he was on good and understanding terms with his soil and environment." The son obviously was given a head start.

And so the book begins with a question from an Oriental proverb: "What hast thou done with the land I loaned thee for a season?" The answer is a record of misuse that began with our ancestors in the ancient world and continues today. Humans, it seems, have always assumed a natural right to the Earth and have ever been optimistic about the limitlessness of its resources—to the bitter end.

Incremental degradation, unnoticed but insidious, brought down Mediterranean cultures through deforestation, watershed destruction and soil erosion. Indications of the same slow trends in North America are extermination of wild animals, decrease in the productivity of fisheries, extravagant cutting of forests, soil deterioration by water and wind erosion, overgrazed rangelands, pollution of water and air and rapid depletion of petroleum and mineral resources. MacEwan holds out the hope that more virtuous living can stop and then reverse the downward slide.

On the bright side, he finds a number of praiseworthy activities whose effect is to restore diversity to the western plains, thereby combatting the monotony of monoculture. Examples are the whooping cranes saved from extinction—at least momentarily, as long as their wintering range on the Gulf Coast lasts—and bluebird boxes that have restored flashes of azure to country fence posts. Reforestation is on the increase, although still too little is being done.

Classed by MacEwan with these unmistakable improvements are others of a dubious nature, oriented more toward solving human problems than to ensuring environmental health. Such are extensions of irrigation to larger acreages, the suppressing of forest fires so that wood mills will not run short, flood control structures at Winnipeg where floods are the normal regime, and an increase in shootable birds thanks to the hunters who fund Ducks Unlimited.

Such inconsistencies—mixing and confusing nature protection with human welfare and economic gain—give the book an interest beyond its factual and historical contents, for in them MacEwan reveals himself and, archetypally, westerners in general. How like him we are in our well-meant conservation programs, even as we continue our quest for sustainable economic growth.

According to the Canadian Encyclopedia, Grant MacEwan was born in Brandon and educated in that city as well as in Melfort, Guelph and at Iowa State University. He was professor and Department Head of Animal Husbandry at the University of Saskatchewan (1928–1946) and then Dean of Agriculture at the University of Manitoba (1946–1951).

Frustrated in his attempt to win a seat in the federal government, he turned to civic government in Calgary where he was alderman in the 1950s and mayor from 1963 to 1966. He became an Alberta MLA in 1955 and leader of the Liberal Party in 1958, resigning from that position in 1960. From 1966 to 1974 he served as Lieutenant Governor of Alberta and began to devote himself to writing, particularly on historical themes.

What is the philosophy of this quintessential westerner? What does he tell us about the character of the prairies, about ourselves?

MacEwan's essays show a preoccupation with two major issues that constitute the tension and dialectic of his thinking. First is his concern for people and their futures, accented by numerous references to caring about grandchildren and duties owed to future generations. Second is his affection for the marvellous world and its non-human inhabitants, especially such threatened innocents as caribou, grizzly bears that "need friends" and prairie dogs of whom he says, "it would be a shame if these likeable little creatures were to lose out completely in their struggle to survive." That the two concerns are in conflict is never granted. On the contrary, his message is that wise resource use, frugal and careful stewardship, will provide the best for both worlds.

His problem is our problem: the conflict between an inherited and largely unexamined value system on which we have been brought up and on which our social institutions are based, and a new value system, just emerging, that challenges much of what we have been taught.

The two belief systems, the anthropocentric and the ecocentric, do not so much pose an either/or choice as a priority choice. Everyone agrees that we people have our just place in the world and that as heterotrophic animals we must use surrounding ecological systems to obtain life's energy and materials. Likewise, a consensus is emerging that the world environment is important; its beauty, diversity and permanence ought not to be destroyed, and we degrade it at our peril. Putting the two together, can we not agree that people of inestimable value exist within an Ecosphere of inestimable value? With that agreement, the important remaining question is, Which comes first? MacEwan, like most of us, has not squarely confronted the issue. The conflict is submerged, and the tension finds expression in contradictory statements.

The book's title, *Entrusted To My Care,* expresses the conventional Judaeo-Christian dogma that humans are special creations, the favoured of God who demands nothing more of them than the careful exploitation we like to call stewardship. This message is reinforced by various statements scattered through the book that imply a world created exclusively for the chosen species.

Resources are for using, MacEwan says. Hoarding would be folly (a message dear to the hearts of those with oil and gas wells). Water belongs to everyone. Trees must rate with the Creator's finest gifts to mankind. The forest is a complex thing and should be used to bring the greatest good to the greatest number. What greater challenge than that of bringing wise and generous guardianship to the natural heritage?

Yet even as he expresses these popular beliefs and recommends stewardship (meaning people firmly in charge), MacEwan is uneasy. It is wrong, he says, to suppose the world was created primarily to serve mankind's purposes and pleasures, and he warns against "unwarranted self-importance, even worship of the self more than the Creator." In his essay on the European conquest of North America, he belittles the idea of divinely sanctioned resource-use though he then dilutes the message by adding, "unless the proceeds are unselfishly shared."

MacEwan is wary of elevating to a place of high value any but the Creator and his finest two-legged invention. He shies away from discussing human population as a problem, a subject in which humanists scent danger because its serious consideration raises the spectre of condemning human fecundity. He gravitates to the easy position of advocating good resource husbandry and wifery that believes, against the odds, that we can save the resource cake and eat it too. With a hope and a prayer, the merciful custodian, caring and not wasting, will assure for humanity the Good Life in a world of diversity and beauty. Abundant evidence indicates the hope is hollow.

As long as we see ourselves as the centre of creation we will rationalize our proclivities to use, waste and destroy whatever parts of the world our technology qualifies as "resources." Only recognition of the world's inherent and exceptional worth can rescue the world from its most pushy creation.

The priority we must begin to seriously consider is Ecosphere before community, ecosystem before organism, the whole before the part. The planet is more than its people.

MacEwan turns to this saving message in a thoughtful "Supplication" at the end. Had he placed his call for a higher degree of Earth-care firmly in its context, the hopes that he expresses for the future of Earth and Humankind all through his book would shine more brightly.

> Remind me repeatedly that there can be no hope of genuine peace on Earth as long as humans are at war with thy other creatures and creations, meaning nonhuman fellows, soil, trees, grass, flowers, water, air, minerals and natural

grandeur, which are placed here as part of thy wise design and in point of wonder far surpass man's most complex inventions. . . . Call it what we will—conservation, concern, preservation, brotherhood, stewardship, or Godliness—grant, oh Master, that as the highest of all goals it will capture more human imagination and breed a purpose that will far exceed the savage lusts and greedy exploits that ravage thy treasures . . . Amen.

FACING REALITY

ECOLOGY AND POPULAR SCIENCE

In a classic paper published in 1935, the British ecologist Arthur C. Tansley discussed the ecological realities of the world:

> Though the organisms may claim our primary interest, when we are trying to think fundamentally we cannot separate them from their special environment with which they form one physical system. It is the systems so formed which from the point of view of the ecologist are the basic units of nature on the face of the Earth.[1]

Tansley coined the term *ecosystem* for these fundamental Earth-surface units whose reality he suggested but made no further attempt to define. The idea is simple yet elusive. We have not yet learned to visualize the Earth spaces in which we live as Living Spaces, as vital surrounding systems that sustain us. Yet when these Living Spaces are endangered so are we.

A primary challenge for everyone is to think fundamentally, to get to the roots of our relationship with the planet, to dig below everyday language and concepts. Tansley's invitation to question the organic parts, to see beyond the bits and pieces that are taken to be wholes, to understand more comprehensive surrounding realities, has become an essential task today. The environmental ills smiting the world, as well as those that wait threateningly in the wings, are not acts of God sent like Job's boils to perplex humanity. Rather they are the result of ignorance, reflecting incomplete and fragmentary concepts about the world and the place of people in it. When conventional knowledge is wrong, wrong ideas and misdirected activities flow from it.

Tansley's summons to identify and sympathetically understand the "basic units of nature on the face of the Earth"—the forests, the grasslands, the lakes and streams, the mountain wildernesses, the farmlands and the settled lands in all their three-dimensional complexity—has not yet been heeded. So far we have failed to comprehend and appreciate the living world as a large global ecological system made up of smaller ecosystems. It is these organism-enveloping ecosystems, not species, that Tansley referred to as "units of nature on the face of the Earth."

What is important today is to change our understanding of the world, to focus on ecosystems rather than on the individual species and organisms that are parts of them. Such changed understanding of surrounding realities will fundamentally affect how we live in our planet Home.

ECOLOGY MISCONCEIVED

To perceive the world whole, to sense the integrity of nature, to understand the parts—atmosphere, seas, rocks, organisms—all playing their appropriate and different roles, while comprehending their underlying unity is our fundamental task. But in many ways, Ecology has missed its calling.

The word "ecology" is derived from the Greek "oikos" which means house or home. Therefore a literal translation of ecology is "the knowledge of home" or "home wisdom." As such, it invites study of the world's living spaces and all that is within them. Unfortunately this insight, this vision, has been blinkered by a focus on organisms. Like the social sciences, ecology's own scope has been limited by a preoccupation with organisms, with protoplasmic bits and pieces, often assumed to play out their roles with little reference to the larger systems that surround them.

The reason for our limited vision is historical. Ecology grew out of natural history, stimulated by the findings of geographers who early in the nineteenth century began to travel widely and observe the variety of plants and animals from the tropics to the poles. Alexander von Humboldt laid the foundations after his famous expedition (1799–1804) to Central and South America where he realized that climate was reflected in vegetation. Soon, other plant geographers such as Schouw and de Candolle were attempting to formulate laws relating the distribution of species to light, moisture and temperature. The word "ecology" entered the language in mid-century (Thoreau used it in a letter in 1858) and the discipline was confirmed in 1895 by the publication of Danish professor Eugenius Warming's influential book translated into English as *The Oecology of Plants.*[2]

Like many recent fields of knowledge, ecology came out of an established science: biology. Even today the assumed best preparatory training is biological. Lumbered with this background, those professing to be ecologists have naturally brought to their studies a fascination with the liveliest components of ecosystems—species, populations and communities—which have, however, distracted attention from the larger realities whose parts they are and where much of their meaning resides. We have not been able to see the hive for the bees nor the forest ecosystem for the trees.

The rooted conviction is that the entities of prime importance on the Earth are plants, animals and especially people, rather than the globe's miraculous life-filled skin. *Species* attract far more attention than the Earth-surface *Spaces* that envelop them, even though, over the long haul, the Species were born from the Earth-circling fertile Space that continues to provide for their renewal, support and sustenance. *Endangered Species* elicit torrents of public concern; *Endangered Spaces* are routinely desecrated and destroyed with scarcely a murmur of public disapproval. The priority is wrong, and from this profound error the whole world suffers.

Like a whooping crane chick hatched by sandhill cranes, ecology has not yet discovered its singularity nor declared its independence. It has not outgrown its lowly initial status as "the fourth field " of biology, the last to arrive after morphology, physiology and taxonomy. As a result, ecology has been conceived as a discipline that plays around the edges of biology rather than as a more comprehensive discipline that integrates biology with all the Earth sciences.

Ecology's task is not peripheral but central, for what is more important than efforts to comprehend the overarching supra-organismic reality that we have called Nature, the World, the Ecosphere? The aim is in the name: to develop an inclusive knowledge of the Ecosphere and its ecosystem parts. When the Earth-home, our Home Place, assumes reality in our minds and imaginations, then new and important ethical concerns will follow.

The sciences need this wider context and framework. Biology by itself is incomplete. Organisms do not stand on their own; they evolve and exist in the context of unified ecological systems that confer those properties called life. Life is not a property of protein molecules nor of protoplasm; it is a property of the ecosystems that the planetary Ecosphere comprises. The panda is a part of the mountain bamboo forest ecosystem and can only be preserved as such. The polar bear is a vital part of the arctic marine ecosystem and will not survive without it. Ducks are creatures born of marshes just as cacti are one with their deserts. Biology without its ecological context is dead.

The tree of knowledge branches into finer and finer divisions. Geology, for example, that began as study of the whole planet, is subdivided and reduced to a multitude of specialties: petrology, lithology, volcanology—each an element of the original subject. Similarly, by reduction, biochemistry has budded off from biology to study organic phenomena at a lower level than whole organisms. Chemistry carries the analytic division even farther, while atomic physics deals with a simplified version of chemistry's universe. In this sense, physics is a degenerative form of chemistry, chemistry a degenerative form of biology.

By the same logic, the studies of organisms and species in biology represents a lower-level focus on subdivisions of the world, on fragmentary parts of larger home-systems or ecosystems. But home is more than the creatures in it, important though they are. Therefore biology, rightly understood, is a degenerative form of ecology, dealing with one component abstracted from ecosystems. Biology is a subdiscipline of ecology, not the reverse.

The time is right to rethink ecology, to understand it properly. Ecology is, or should be, the study of ecological systems that are home to organisms at the surface of the Earth. From this larger-than-organic perspective, ecology's concerns are with volumes of Earth-space, each consisting of an atmospheric layer lying on an Earth/water layer with organisms sandwiched at the solar-energized interfaces. These three-dimensional air/organisms/soil-water systems are the Real Thing. Furthermore, Earth-surface ecosystems are even more beautiful, more complex, more important, more deserving of attention and ethical concern than the organisms contained as parts in them, however attractive and fascinating they may be.

THE ROOTS OF MISCONCEIVED REALITY

Before the modern era, belief was widespread in the existence of universal orders of organization, surpassing in importance organisms in their populations and communities. The concept of all nature as an organized whole informed by reason was central to mainstream Greek natural science from Plato to the Stoics, and it carried over to such famous Romans as Cicero and, much later, to Leonardo da Vinci.[3] Although gradually submerged in the Middle Ages, the idea persisted and was implicit in the counter-culture thinking of nineteenth-century Romantics both in Europe and North America, contributing to the philosophical framework from which concerns for conservation developed. Influential in subtle ways over the last four hundred years, it has nonetheless been viewed as out on the radical edge rather than within mainstream thought.

The change in thinking that disintegrated organic nature and severed the Chosen Species' roots in nature is a legacy of the Renaissance and the Age of Enlightenment. Back then, nature was divested of mind and soul and rendered at once dead and menacing.[4] The stage was set for a mechanical and materialistic view of the world consistent with the technology of labour-saving machines and clocks that developed rapidly in the middle ages. Gradually, mechanical models became the symbols by which people understood themselves and their surroundings. The enchanted world receded as the masculine sciences took centre stage.

God was accommodated and dismissed as the Prime Mover, the clock-maker who wound up the clock of the universe, set it ticking, and then disappeared, satisfied with a job well done. Because the Judaeo-Christian God is a God of history rather than of this world, His presumed exit was interpreted as a signal for man to take charge. In the 17th century, Descartes gave explicit form to the universe-as-mechanism idea, the key to manipulation and control. Since then, the West has practiced Cartesian science, studying to discover what makes the material world tick, searching within matter for accessible levers of power. The purpose of science, said Francis Bacon, is to control Nature and force her to do humanity's bidding.

A clock-maker fabricates his timepieces from gears and springs that precede the clock and apparently explain its workings. Even when the clock is put together as a whole, the understanding of "how it works" is a function of its parts and their movements. For machines, reduction is the key to comprehension.[5] Applied to the world of nature, everything is to be understood by composition, by component parts, smaller and smaller. The physicist studying subatomic particles is closest to the Truth. Such a mindset creates a fragmented world. Further, it insinuates into our minds the mischievous idea that the fragments, the parts, are somehow more important than wholes.

The mechanistic worldview delivers great power. Cartesian science has proved successful in providing knowledge-for-control in such fields as physics, molecular biology and medicine. The procedure has the appearance of being totally effective because it follows the path of least resistance, pursuing problems that yield to it while bypassing those with which it cannot cope. Disciplines where it falters or fails—psychology, anthropology, sociology, evolution, neurobiology, embryology, ecology—are written off as non-science, as "stamp-collecting," to use Lord Rutherford's phrase for all non-physics studies. "Successful scientists," said Nobel laureate Peter Medawar, "tackle only problems that successfully yield to their methodology."[6]

Just because reductionism has delivered power in certain fields does not mean, however, that it opens the window on reality, that the actuality of the world is to be found in its parts, that Truth resides in atomic and subatomic particles. The frequently asked question, "Can the whole be greater than the sum of its parts?" gives the game away, for implicit is a prior commitment to the parts. In effect, the question says *we know* that the parts exist, now what about their sum? Thus the rightness of reduction is *assumed* by questioning whether anything other than parts can really exist.

Carried over into society, mechanistic reductionism tracks the cause of tuberculosis to a bacillus rather than to slum housing, the cause of cancer to oncogenes rather than to industrial pollution, the cause of evolution to genic mutations rather than to co-development with larger surrounding systems.[5] Science does not entertain the awkward possibility that reality might be distorted by giving priority to parts over wholes.

Books and tracts abound explaining that the individual is more important than the social group, the person more important than the world that encapsulates her, the fetus more important than the woman that encapsulates it. Any organism, we are told, can be computed from the complete sequence of its DNA. The brain is a holograph, the body a machine. How else do we explain the success of bioengineers in replacing the grit, glue, jelly and soup of the human body with neater and more efficient metals, plastics, ceramics and semiconductors?[7]

In short, the Cartesian heritage is a fragmented perspective, focused downward rather than upward. The search for meaning at lower and lower levels of organization blunts the higher-level search for more inclusive realities. One-way vision threatens the future of the human race by blinding it to the surpassing importance of supra-organismic realities—the Earth's sustaining ecosystems, the planet's skin, the Ecosphere.

ETHICS AND THE ECOSPHERE

Perhaps the greatest mischief of scientific materialism and explanation-by-reduction is what it does to ethical concerns. By conceiving things mechanistically and shifting meaning to their parts, modern science strips away all sense of intrinsic value.[8] It destroys the intuition that things can have importance for their own sakes, independent of their parts. What real empathy can be felt for a machine-dead universe whose explanations reside in its atoms?

A meaningless universe leaves little in the conscience of people but a sense of their own diminishing importance. The last religion left to them is a slowly evaporating Humanism that isolates them from the world of nature and leaves them alone, clinging to each other in a prison of their own making, bravely repeating that only they are important, only they have souls, only they will reap rewards in the Great Hereafter. "Human rights" is the leading secular slogan.

From the precept that only humans matter, a disastrous corollary follows: The world is for exploiting. Parks are for people, animals are for shooting, forests are for logging, soils are for mining. The sole basis for ethical action is the greatest good for the greatest number of people. The values of all things lie only in their abilities to serve us.

Contemporary morality—the sense of right and wrong—is completely in-turned, completely focused on humankind. That focus makes it difficult to be sensitively concerned about the world in the face of escalating human demands. Sustainable development, we are told, must include forceful economic growth, for how else can the needs of all the world's people be met?

Lacking an ethic that attaches importance to *all* surrounding creation, people continue to do the wrong things for the apparent "good of humanity." *People First.* 6.2 billion people going for 10, all believing in *People First,* increasing their wants without limit, are a sure recipe for species suicide.

If what is wrong is to see the world in pieces, fragmented into atoms, species, resources, with people alone important, then what is right is to re-perceive the world as one, a whole, organically complex, beautiful beyond compare, and to reorient to it in ways that confer first importance on it. In short, replace the homocentric with the ecocentric viewpoint. Then the intrinsic values of the Ecosphere and its living realm will be recognized, as will the rights of things other than human to exist in and for themselves.

Once values are straight, everything else can fall into place. An ecocentric worldview, valuing the spinning planetary home above the organisms hitching a ride on it, elevates in importance the ecosystems that humans call land. Love of the land, love of place, love of our endangered Living Spaces, is the grassroots cure for the sin of species narcissism.

The world was not created for people only, but for purposes that transcend the human race with its limited foresight and imagination; therefore it behooves all conscious inhabitants of this superb planet to nurture it as a garden, maintaining it in health, beauty and diversity for whatever glorious future its denizens may together share.

CHANGING THE GLOBAL VISION

An early creation myth, at least 4,500 years old and a strong influence on the Hebraic legend of Genesis, comes down to us from the Babylonian civilization that prospered on the rich alluvial floodplain of the Euphrates River in Mesopotamia, now Iraq.

In the beginning, so the story goes, river met sea in a misty estuary whose intermingled clouds, fresh water and salt water gave birth to fertile silt, from which in turn sprang the gods of Earth, sky and horizon.[1] Thus from water came all other material things of the Earth, with life and soil and finally humus-beings, for "human" and "humus" are derived from the same clay root.

The maker of this myth, forerunner of the sedimentation geologist, clearly understood the importance of deposition by moving water. Attracted and captivated by its generative power, she merits recognition as the first alluvial fan.

Earth scientists are today rephrasing the creation story, tracing the genesis of the watery world with its air and soils to dynamic forces that still continue under a flaring fire-storm sun—wobbles in the axis of the spinning planet, magnetic wanderings, meteorite impacts, volcanic outpourings, crustal plate migrations, glaciations, energy and material exchanges between sea and air and, accompanying them all, the pervasive influence of organisms that apparently were present in microscopic form on Earth from the beginning.

A quantum leap in organic influences accompanied the quite recent rise to power of the Featherless Biped, first mastering the uses of tools and fire, then the culture of animals and plants and finally the control of fossil fuel energy for industrial purposes. In an uncomprehending way, people have become a potent geomorphic force that is rapidly and perilously changing the face of the Earth.

Not many years ago the world seemed steady and secure, a dependable platform whereon the cycles of day and night, new moon and full moon, spring and fall, life and death, were repetitively dramatized in tune with the central human pageant. "All the world's a stage," wrote Shakespeare, "and all the men and women merely players."

Today's appreciation of the world in its cosmic setting suggests that the stage itself is creatively evolving, ever-changing, in unison with the organisms on and in it. The world is a leading performer, a star of the show, and not just a decorative backdrop that casts in relief humanity's role. The play goes on and on, marvellous and surprising, continuously written and rewritten by the genius of all its participants.

The physical and biological sciences are revealing the vital vigour of our planet and the interrelationships of all its parts. At the same time the unexpected effects of technologies that these same sciences have generated are forcing a new consciousness of the intrinsic worth of the beauty and freshness of the natural Earth. Through the use of science/technology, human enterprise has enriched us and contributed to our welfare, but with high costs to the Earth-source, costs that can no longer be deferred.

Here is the dilemma: physical and biological knowledge—essential for our sympathetic understanding of the surrounding world—also provides the means by which it is polluted, degraded and threatened with total destruction. The misused power of science underlies the human-induced global problems that during the last decade have turned international attention to environmental protection, sustainable development and conservation planning for resource use on land and in water. The intent of all such strategies is to ensure the continued prosperity of the human race by "green" growth and development, doing what we have always done but more carefully, cleaning up pollution as fast as we make it. Yet that way, we will lose the game.

At the moment, no clear resolution of our predicament is in sight. Radical changes in culture follow new radical ideas, new insights, new beliefs. The solution to the present quandary is not beer-can recycling, back-yard composting and planting trees to soak up CO_2, useful though these interim measures may be. The solution is a deeper and more profound leading vision as to what "environment" really is, enlightened by affection for the place that is home to us, guided by a more comprehensive evaluation of ourselves, of where we are and of how we came here. We need the insights and skills of poets and philosophers to make the vision compelling. In all new beginnings is the Word.

NEEDED: A WHOLE-EARTH VIEW

Throughout history we have always seen ourselves and our surroundings through lenses fashioned by past experiences and past beliefs. What we already know filters and focuses what we perceive. Parents and teachers outfit each youth with cultural belief-spectacles that direct and colour a particular world view. Once formed, it becomes both self-evident and "right" for those born to it. Then the world views of other cultures seem strange and incomprehensible; we judge them to be primitive or inane compared to our own true perspective.

When, as today, signals from the Real World clash with the expectations of the world view constructed by our particular culture, then the time is right for a new look, a change to better glasses, keener insight, improved theories about where we are, who we are and what we should be doing.

Already we can envision, in outline, some parts of the emerging new look, for now we know the planet as a beautiful floating cell that supports people as self-conscious parts of its fertile skin. Could our sense of time be collapsed from Earth's beginning to today, we would see in an instant our intimate relationships to air, water, soil and all other organisms. These also are parts of the marvellous Earth-system, companions on an evolutionary journey whose end is unknown.

Could our spatial relationships here on Earth also be seen instantaneously, we would recognize the ecological dependencies that make foolish our attempts to dominate, control, manage and reshape the Earth-environment in the immediate interests of one single species. Waging war on our environment for short-term gain brings long-term pain.

Today's appropriate *Weltanschauung* is the word translated with a twist. We do not need another "world view" but a new "view of the world," an outside perspective that reveals the Earth in a way that is truer and brighter, more vivid and more accurate than we formerly possessed. Already we have glimpsed a surrounding reality that is whole, a short step from holy, a world that lays claim to our loyalty because we are dependent parts of it.

This vision challenges the assumption that we are justified in attempting to supervise and exploit all circumstances for limited human goals. It calls into question the widespread masculine belief that the leading purposes of thought and action are to master and manage. Although it turns our traditional relationship to the world upside down, humanity's future rides on just this authentic realization.

SIGHT BECLOUDS VISION

Conflicting ideas are vying for society's attention. Many pages of our newspapers are devoted to stories of environmental deterioration due to industrial developments. But turn to the business sections and find hand-wringing reports decrying the slowness of economic growth. We read that we ought to act more altruistically toward the poor and the powerless, toward the developing countries and the environment, yet individualism and privatization are elevated to the status of supreme civic virtues. While thoughtful people suggest that the progressive road into the future is co-operation, the competitive route is far more popular in the world of real politics.

Some trace the cause of our dilemma to the education system, and certainly it is geared to competition in every possible way. But education is a cultural pursuit, reflecting cultural norms; wherever the latter lead, it follows. Others blame the scientific framework provided by Darwinism, but his ideas were derived from Adam Smith and Malthus. The competitive model that Darwin adopted long preceded him. Perhaps its roots are in the very way reality is perceived in Western culture.

The senses by which we experience our environments are unequally developed. As people, we are primarily seeing animals, our intellectual activities closely tied to vision. Even in sleep we dream in the language of visual metaphors. "I see" is a common synonym for "I understand." The hearing, tasting and smelling senses—their specialized receptors close to the eyes—support sight. Touch, the fifth sense, is different, more generalized, with receptors over all the body surface. Unlike sight, touch requires nearness, contact, intimacy—appropriate to emotional response as indicated by such phrases as "I am touched" and "My feelings are hurt."

McLuhan thought the visual bias to be artificial, a learned response to the technology of the printed word that moved civilization away from an earlier aural-tactile tradition, substituting "an eye for an ear." The senses are out of balance, he said, with sight suppressing the other senses in injurious ways. The eye is cool and neutral, and in a highly visual society all people are alienated, outsiders looking in.[2]

Is the eye necessarily "cool and neutral?" Certainly the feelings aroused by sight can be diluted by distance. The eye senses remotely and, to that extent, is out of touch. But seeing has its different modes, influenced not only by the position of the viewer in relation to what is seen but also according to the sympathy that the viewer brings to active perception. Outlook is "cool and neutral," but insight is warm and involved.

To look *out* as opposed to looking *within* is to see things out of context, separated from the matrix in which they are embedded and to which they are related in space and in time. That is the way we see the Ecosphere, as fragmentary and made up of all sorts of separate things, because we are inside it looking out and around. Knowledge based on vision-turned-outward tends to be detached and alienating, divorced from sympathy, compassion and other emotions that awareness of contact engenders. Such "I see" knowledge unleavened by "I feel" knowledge can be dangerous knowledge.

Sight-knowledge from the inside out wears the face of disinterest. It seems to be neutral, unbiased and aloofly rational, and these supposed characteristics have been elevated to virtues that are then invoked to justify the I-it environmental relationship that projects an impersonal world of separate things: the universe of science.

The objectivity of dispassionate science is a dogma widely accepted by the public and even endorsed by some scientists—when they take time away from emotional defenses of their pet theories and hypotheses. But in truth philosophers of science have rejected the separation of viewer and viewed, of subject and object, of knower and known, that in earlier times was considered the essence of the scientific method. "All my knowledge of the world, even my scientific knowledge, is gained from my own particular point of view, or from some experience of the world without which the symbols of science would be meaningless," said Merleau-Ponty.[4] Science's presumed objectivity turns out to be a subjective belief, confirmed as a liability, an obstacle, a deterrent to sympathetic knowing in the full sense that comprehension implies.

The fact that perception is an active process shaped by beliefs and values and not a passive camera-like response as McLuhan's theory suggests, means that sight need not be alienating. Nevertheless, the apparent separateness of the objects that we see in nature is encouraged by the inside-out perspective. One-way sight invites a faulty outlook that sets up the visual environment for reduction; when uninformed by knowledge of the whole, it disintegrates and particularizes. The antidote is the outside-in perspective, for "looking into things" discloses attachments, bonds, relationships. Out-look disconnects while in-sight, like touch and hearing, connects. In-sight is high-order intelligence.

CORRECTING THE INSIDE-OUT PERSPECTIVE

Today's erroneous World View is the result of inside-out perceptions. The phrase is appropriate, for in common speech it means "wrong." Immersed in the world ecosystem, we have not grasped the meaning of our true environment. Lacking the outside-in perspective, we have misconceived, in a fragmented way, our circumstances, our surroundings and constructed fields of knowledge, disciplines, educational systems, departments—an entire culture of arts and sciences—around the fragments.

Now the revelation of the environmental whole from outer space, interpreted by ecological understanding, is challenging the central ideas of human pre-eminence and purpose that have brought the world to the brink.

The unity is the Ecosphere, the global "being" whose inseparable physical/biological parts have evolved together for 4.6 billion years. The Ecosphere is not an organism nor even a super-organism. Its properties and complexities lie at a higher level than the organisms, air, water, rocks and sediments that it contains. It is best described as *supra-organismic,* meaning above and other-than the organisms that it envelops. Life is a miraculous product of the Ecosphere, one of its important characteristics. Life is not an isolated-in-a-hostile-world phenomenon.

Homo sapiens, a single member among 20 million fellow creatures, is an active component, reliant on the whole. This realization provides a rational foundation for recasting humanity's role as a compliant element within the evolving Ecosphere and the various land and water ecosystems that it comprises. A choice is offered: Are we to be a knowing cooperative part of the whole or a selfish scourge? Will we be gardeners of the world or a blight on it?

A SYMPATHETIC SCIENCE

Science which was first an exploration of God's design revealed in nature, then a means for human mastery and control over nature, has inadvertently opened a new chapter in the book of knowledge whose dazzling insights can change fundamental ideas about the planet-people relationship, about values and about the way the world is known. The re-conceptions of reality, of what is centrally important beyond human needs and wants, opens avenues of escape from both tradition's self-serving mode-of-knowing and its destructive species-centered ethic.

Science may indeed recapture its earlier promise as the instrument of human self-enlightenment. Science will fulfill its high purpose by contributing to self-awareness, assisting the race to feel its way toward becoming more truly human.

No one can say with certainty what "becoming more truly human" means. Aldous Huxley called the shared insight of the world's great religions the "Perennial Philosophy."[5] and it suggests, as a beginning, the achievement of a perspective outside ourselves, a shedding of species selfishness and self-infatuation, a vision of humanity in the context of things more important, "under the aspect of eternity" as philosophers have phrased it.

To this end, questions of human identity and awareness are central. Who people believe they are in relation to the rest of creation and what they believe to be important in their lives will channel their actions constructively or destructively within this one and only planetary Home.

Who in the World do you think you are? What on Earth are you doing?

TECHNOLOGY AND ECOLOGY

Technology has become the primary means by which humans interact with the Earth-home. An outcome of artifice guided by human beliefs and purposes, technology makes visible the deep beliefs that consciously and subconsciously motivate society. Through it the people/planet relationship is made explicit. Hence the importance of a critical appraisal of technology, of its effects on the world and of the extent to which its various forms are appropriate.

Technology, the Big T, has been variously defined. All agree that it is more than hardware, more than machines, tools, material instruments. An inclusive definition of T is: *a reproducible and publicly communicable way of doing things*. The key word *communicable* shifts T into the world of ideas, language, beliefs, culture—which explains why T-transfer from industrialized to non-industrialized nations is difficult until, either directly by acceptance or indirectly by acculturation, the receiving populations adopt Western values, beliefs, perceptions. To benefit from our technology, they must first believe what we believe and want what we want. According to a recent Mercedes-Benz advertisement: "Technology is only opportunity knocking—the secret is to open the right doors." To be effective, T must be directed to the right openings, the right visions, such as owner pride and smooth high speeds in the Mercedes model M-B 300.

An essential part of T consists, therefore, of cultural ideas, especially the values and goals espoused by influential people—in our society by business people. Their materialist philosophy provides the inspiration and encouragement for industry's material production. The built environment, steadily expanding, is the visible expression of T's cultural authority, influence and growth. Progress is T, and T is progress.

TECHNOLOGY AS PROGRESS

Faith in T is deep rooted. Aldous Huxley considered it the modern religion, its symbol a decapitated cross and the Model T Ford. T is our provider and in it we trust to give us our daily bread and deliver us from evil. Yet the Good Life so provided is not free. Environment has been picking up the tab, paying our way. Now its ability to support humanity's appetites and wastes is running out.

The conventional use of the word "progress," often preceded by a resigned or cynical "you can't stop it," brings to mind T's gifts to our consumer society: living better electrically, acquiring more machines, eating exotic foods, travelling faster and perhaps some day visiting the moon and planets. Progress is living "high on the hog" for the largest possible percentage of the human population, using more and more energy and processing more and more materials from the Earth's crust. Will we continue to call it progress as it slowly kills us?

Progress in a finite ecosystem cannot mean the absolute ascendancy of one species. Progress must mean the achieving of a creative symbiosis within the Home Place, where sympathy and care are extended by the dominant species to the rest of creation To transcend traditional preoccupations with our own kind, preparing to appreciate the Ecosphere with all that is in it, requires an understanding of T and how we have used it both to exploit and to distance ourselves from Mother Earth.

EXOSOMATIC TECHNOLOGY—APPENDAGES FOR PEOPLE

Technology churns out a variety of artifacts: tools such as pile-drivers, media such as television, cocoons such as buildings and cars, all of which like spectacles and hearing aids can be hooked on the body to increase the wearer's abilities and pleasures. Aircraft are fast legs, attached by seat belts to people in order to increase their speed of travel. Submarines are worn to swim under the oceans, microscopes and telescopes to see better and farther, telecommunication systems (giant vocal chords) for long-distance information exchange, bulldozer and dragline are muscles for Earth moving, computers extend the mind's scope. Transmitted from generation to generation by culture rather than by genes, T is the new means of human evolution, increasingly adding to human versatility, power and size.

From this perspective, T provides (in addition to biomedical inside-the-body gadgets) a wide range of outside-the-body or exosomatic instruments, extending human abilities to change the world and all that is in it. Side effects are extraction of vast quantities of "raw materials" from the planet's surface to manufacture the array of technologic products, and a return flow of wastes and poisons at all stages from primary production to obsolescence.

Initially purpose shapes the instruments, but soon, like body parts, they begin to inspire their own use. Just as those born with good vocal chords can with difficulty be dissuaded from singing, so a boy with a new hatchet finds much that needs chopping. The United Nations dictum, "Wars begin in the minds of men" is a half-truth. Wars are also encouraged by the availability of arsenals of shiny potent weapons whose purpose is destruction. Disarmament is effective because it amputates the striking arms, separates the muscle-toys from the aggressive boys, removes the temptation to let the exosomatic instruments "do their thing" at the slightest provocation.

If technological instruments are body extensions, may they not also be viewed as disfiguring outgrowths, as excrescences with pathological tendencies? Three potential afflictions we suffer from "wearing" our technological armour are gigantism, addiction and alienation.

TECHNOLOGY AND GIGANTISM

All technologies whose goals are increased power and control effectively lead to gigantism. Machines behave like enlarged limbs and organs, demanding energy for growth, repair and reproduction just as organisms do. A North American, considered together with her exosomatic appendages, is eighty times the size of a Bangladesh peasant judged by appetite for non-renewable resources (fossil energy, minerals) and renewable (ecosystem) resources.

Each Western person bestrides the world like a Colossus, leaving his Sasquatch imprints on land and water through the massive extraction, consumption and waste of resources that enlarged body size necessitates. Head counts are misleading when technologic size is not factored in. Is Canada underpopulated with 36 million people? Before answering, bring in the technologic multiplier, the energy use per capita, that indicates our impacts on environment. Considering that every Canadian stomps the Earth like eighty peasants, our country is vastly overpopulated. Further, in a world of rapid population growth but fixed space, of polluted renewable resources and dwindling non-renewable fossil fuels, the expectation that everyone can be an exosomatic giant on the North American model is unrealistic. The dream of the world's poor, who make up the majority of the human race, casts over the future a dark cloud that will only be dispersed when the rich nations begin to share by divvying up their wealth. Sharing is more realistic than depending on economic growth, for the world will never support, at Canadian standards, the seven billion people expected in the year 2008. One billion could perhaps live as well as the citizens of Calgary and Winnipeg. Some say only 250 million.

The gigantism that industrial T confers is largely material. It effectively extends the power of our material bodies without doing much for our immaterial minds. Like a growing brontosaur, T's body size has far outdistanced brain size. Encouraged to be consumers, we surround ourselves with more and more artifacts, increasing our appendages and faculties, while our reasoning power and modicum of wisdom stay the same or shrink before the TV set. The distortion of the body/brain ratio creates feelings of helplessness: The common fear that machines are out of control, that they are running people's lives, suggests that exosomatic evolution has outstripped its rational management.

If the diagnosis is correct, we would do ourselves a favour by reducing exosomatic T to restore a better body/brain ratio. In industry, fewer mega-projects and more "soft" local developments would reduce our gigantism. In agriculture, a larger eye-to-acres ratio would improve on the wheat-belt trend of fewer eyes on larger farms.[1] Instead, the favoured prescription is to keep T growing, not attempting to enlarge comprehension per se but to supplement and beef up brain power with the latest T: AI, Artificial Intelligence, that also doubles for Artificial Insemination. If only T could deliver AW, Artificial Wisdom!

TECHNOLOGY AS ADDICTION

Because they give us power and pleasure, exosomatic instruments are addictive. When "worn" even for a short time they become necessities, difficult to lay aside unless replaced with better, more efficient models. The rancher turns in his horse for a half-ton truck. The youth moves up from motorcycle to muscle-car. For those habituated to fast movement through the air, surface travel—except perhaps for recreation—loses its charm.

Progress is traditionally conceived of as greater power, control and efficiency—not less. To suggest giving up powerful Ts on the suspicion that they are dangerous and destructive is to invite scorn and accusation of wanting "to go back to the cave." Georgescu-Roegen thought that this addiction—as attractive to humanity as a flame to moths—is the greatest obstacle to a rational human ecology. He was pessimistic about cures for the pervasive dependence on exosomatic comforts. "Perhaps the destiny of man is to have a short, but fiery, exciting and extravagant life."[2]

Like addictions to drugs, sex and gambling, being hooked on T is controllable if not curable. The first step is recognition that we have a problem, that T is a predicament, which then suggests the need to understand it and find satisfying alternatives to its destructive facets. The sometimes brilliant philosopher William James thought that the moral equivalent to warfare was an all-out battle with Nature, a constructive alternative to killing each other! We have proved that such battle is an immoral equivalent, for it kills everything. Perhaps a moral equivalent to war can be found by distinguishing friend from enemy, separating appropriate T that assists our symbiosis with the universe from the poisonous and destructive T of heavy industry and military might, then launching an all-out attack on the deadly kind before it destroys the world and ourselves completely. Let us declare our debilitating addiction to instruments of power and explore the possibilities of cutting Big T down to little t.

TECHNOLOGY AS BARRIER

T is also an alienating influence, interposed between people and the planet. As the manufactured milieu grows, contact between people and the Ecosphere is steadily reduced. The built environment acts as a filter and barrier that "progress" renders thicker and more opaque, gradually eliminating and shutting off nature's direct sensual stimuli. During the last eclipse of the sun over southern Saskatchewan, children in many schools were only allowed to view that celestial phenomenon on television. Watching TV in buildings without windows is the model of self-induced alienation, annihilating all sense of ecological roots and dependencies.

T has been described as arranging the world so as to minimize direct experience of it, an alienating effect that results in sensory deprivation. Cut off from Nature's sights and sounds, people end up in single-species confinement. Aloneness in the city, with nothing to sense but themselves and made-things, induces various psychoses. T's cages isolate people from the Ecosphere milieu—their biological and evolutionary home—as effectively as iron cages isolate lions from their savannas. Just as zoo animals deteriorate when deprived of their natural home surroundings, so also humans as Earthlings deteriorate when, like the legendary Antaeus, they lose touch with planet Earth.

TECHNOLOGY AS PLACENTA

From the standpoint of human ecology, T is the system of values, beliefs and techniques by which a society, a culture, taps into, uses and modifies the Ecosphere, in the process changing both itself and the surrounding system. T is the instrument of contact between people and the miraculous enveloping environment from which over four and one-half billion years of evolution they came. T is the means by which humanity extracts what it wants from the Earth and returns what it does not want to the Earth, playing an analogous role to the placenta and umbilical cord connecting fetus and mother. Should the fetus imperil its host by excessive demands and generation of wastes (a likely effect of gigantism), then both will suffer. As with mother and fetus, the only safe and sustainable ecological relationship between people and planet is one of moderate demands in a symbiotic alliance.

The dependency of people on the Ecosphere is complete. Indeed, all organisms are inseparable from their environments except in thought, and life is more an attribute of the Ecosphere than of the organisms it encapsulates. The natural world and all forms of life within it are interpenetrated. Sophisticated T does nothing to reduce the people-planet dependency, but it does magnify the possibilities for exploiting and poisoning the relationship. The onus is on humanity to develop a technology appropriate to the well-being of the Earth, a T that does not imperil the Ecosphere but contributes to its healthy functioning.

Conventional T is purposive and aggressive. Comprising values and ideas as well as tools, it urges its own uses which, given our cultural past, are people-serving in a short-sighted way. Unless we recognize this fact, we slide into the popular "technology is neutral" mode of thinking that backs away from any kind of control over our inventions and shrugs off the need for ethical choices at T's leading edge. The ultimate nonsense is justification of war because of the technologic advances it brings—poison gases that later yield insecticides, tanks that are the models for better tractors. Danger lurks in the idea that nothing is wrong with any T, that it is only what we do with it, what we make of it, that can be judged ethically. This, said McLuhan, is the numb stance of the technological idiot.

EVALUATING TECHNOLOGY

Many human problems are unsolvable in the humanistic context. What is more important, mother or fetus? If human life, actual and potential, is believed to be paramount, no conclusive answer is possible. But in the context of the Ecosphere, knowing what we know about carrying capacity and of social and environmental destruction due to overpopulation, most ecologists will come down on the side of the mother.

Similarly with T. What is appropriate and what inappropriate cannot easily be decided if the arguments consider only help or hindrance to people. Will it feed more of them, will it prolong their lives, will it make them more powerful, will it help them calculate faster? The usual conclusion is yes, do it, because people will be "better off."

A sounder standard for judgement is ecological, measured against the requirements for symbiotic survival in the Ecosphere. Then the measure of T's goodness or badness is its effects on the Earth system from which it draws materials and to which it returns wastes. Hitherto progress has meant more T, more industrial growth, more material goods, more wealth at the expense of the Ecosphere. Progress must be redefined as sustainable ecological relationships between humans and the Ecosphere, to which new kinds of T evaluated in the context of health and permanency can contribute. That is, "T for Two."

Appropriate T will respect the vital milieu of air-water-soil-sediment-organisms. It will not appropriate these Earth-surface resources faster than they are replenished and renewed. Nor will it introduce into the life-space pollutants and toxins from underground. Of the latter, in particular, it will not play around with radioactive substances nor falsely promote them as safe, clean and cheap. Rather than encouraging nuclear indigestion, T will help the world avoid atomic ache.

NATURE, SELF, AND ART

"I strove with none, for none was worth my strife," wrote that combative old curmudgeon Walter Savage Landor on the occasion of his 75th birthday, and then (one supposes), recovering his sense of veracity, "Nature I loved, and, next to Nature, Art."[1] Here, in few words, are the lineaments of the Western credo: belief in a reality that consists of an external Nature to be loved (or as often hated) by an I-me—in this case a poet—who creates a world of Art.

Landor's Nature-I-Art provides a foil for today's ecological *Ars Perspectiva*. Just as the discovery of linear perspective gave flat-surface painting the depth of vision, so ecology's curvilinear perspective suggests a truer and deeper representation of nature, art and people.

The philosophic worldview to which the West is heir rests on belief in separate things, as exemplified in the threesome Nature, I and Art. Yet when these things are closely examined, a network of binding connections and interrelationships is discovered; the trinity is one. The taxonomy that separates and identifies separate things is, perversely, humanity's own creation. We have voluntarily cut ourselves off from reality, sentenced ourselves to solitary confinement and thereby induced a form of madness.

Descartes was intent on establishing the separate self. His famous statement—"I think, therefore I am"—is a simple equating of self-consciousness with the detached ego. But what about the unself-conscious or instinctive part of the psyche, the Jungian collective inheritance from our biological past? The "wisdom of the body" argues with equal force that the "I" alone (as well as the aggregate I's of society) is suspect as a self-standing structure. The liberal doctrine of individualism that dominates belief in our culture is true only by consensus. The separate self is a proposition to be accepted not by logic but by faith.

Nor does logic support a separate Nature. Humanity is clearly issue of the Earth, physically from Nature's water, soil, air, sunlight and plants and psychologically from Nature's animal kingdom by some generative miracle. The fiction of Nature-as-other, set against humanity, can only be maintained with eyes firmly closed, denying ecological realities, concentrating hard on the fiction of the I-ego and its supreme importance.

Then there is Art, bridge or barrier between Self and the Other, creating new symbolic realities and reshaping the stuff of Nature, "closer to the heart's desire." At first glance it affirms the reality of self, for is not Art the artist's own creation? Partly, yes, but where does the creativity come from, where are the springs of imagination? Inspiration—irrational and unexplainable—arises less from the conscious ego than from deep uncontrolled sources which, baffled, we call "creative human nature." Back to nature for creativity! Of course, the artist's artistry also reflects beliefs that filter and channel the expressions of creativity, encouraging artistic expression to confirm its own autonomy (as in Art for Art's Sake) or to deny it (as in Art for Connectivity's Sake). The artist's vision of Nature and of Self may be in-turned or out-turned, loving or hating, fragmented or whole.

Elaine de Kooning, wife of artist-painter Willem de Kooning, recounted a summer night experience with her husband in the southern Appalachians in 1948. "After ten minutes on the moonlit road, none of the lights from the school were visible to interfere with the vast, heavy, velvety blackness of the sky, nor did sounds of laughter and music penetrate the almost terrifying hush. We stood still, enveloped by the awesome multiplicity of stars. 'Let' s get back to the party' said Bill. 'The universe gives me the creeps.'"[2]

Most of us feel timorous when away from the crowd, uneasy with the aloneness sensed in a universe we have been taught to mistrust. Yet the estrangement from Nature that worries the separate self, gives it the creeps and eventuates in fragmented art forms, is the invention of our own Western civilization. If "little we see in Nature that is ours," the reason is a belief system that requires discrete and separate things in order to support a particular worldview. An unhuman spiritless world of Nature is necessary to sustain belief in autonomous people and also to justify both despoliation of that separate other-world and replacement of it with the practical artistry of a consuming technology: the metaphor of ourselves.

Before consciousness of self developed, before the breakdown of the bicameral mind, Nature-as-other was not conceived. In those a-historic times, Nature without self-conscious parts was all, just as today (so it seems) plants and animals and little children exist innocently in the state of Nature. Consciousness, though an inevitable way of knowing, need not be arrogant. Yet somewhere along the evolutionary way a certain kind of consciousness found it expedient to declare itself not only separate but different in kind and superior to the Other that miraculously gave it birth. The Western ego entered its adolescence, forswearing its parentage.

The resulting alienation reveals itself in uneasiness, in a restless search for roots in the past and for directions to the future to give meaning to the here and now. Reestablishing ties with the ancient dwelling place, the Home Place, is a largely unrecognized but widespread preoccupation. "Getting back to Nature" or better, growing by "Going forward to Nature," has level upon level of meaning that runs the gamut from the most superficial exercises in recreation to the most profound explorations of re-Creation.

The function of art according to its etymology is to join, just as that of religion, itself an art-form, is to bind together again. But joining and binding—the means of restoring wholeness—lose their meaning when fragments and parts are misidentified as wholes. If reality actually consists of separate things, what is the point of skills in joining and binding together? Art forms, for fully autonomous people, lose much of their meaning. Shorn of their deepest significance, they are liable to degenerate into superficiality: religion reduced to ritual, science to power trips, drama to titillation of the emotions, painting and sculpture and music and poetry to baubles and playthings. Bereft of meaning, save as a rich storehouse of resources and metaphors, the world of Nature ceases to influence the practical artisans of the built-world. They proceed to engraft on the planet's living skin an un-organic architectural art that Earth's immune system must sooner or later reject.

In every generation over the last four hundred years, a few seers have questioned the sanity of the self-serving taxonomy that permeates the Western tradition. In effect, such people deny the axioms of autonomous self, autonomous Nature, autonomous art. Mystics seek a not-I completeness by renouncing the ego, guided by the perennial philosophy's fundamental axiom, Thou art not thou, Thou art That.[3] Less ambitious mortals seek peak experiences—those momentary psychic highs when the boundary between self and other, between subject and object dissolves. The same quest for wholeness, metaphorically played out as variations on the theme of Paradise Lost and Paradise Regained, is the heart of great art, according to Northrop Frye.[4] Its attraction suggests a weakness in the duality—the conscious "I" opposed to nature—that appears in much philosophy and reaches its conclusion in Nietzsche's glorification of Darwinian competition as the path to human salvation.

The philosophic tradition that began with Plato apparently ends with the Superman, whose motivation is his will to power and whose goals are whatever the aristocratic personality dictates. "Life," said Nietzsche, "is essentially appropriation, injury, overpowering of what is alien and weaker; suppression, hardness, imposition of one's own forms, incorporation and at least, at its mildest, exploitation."[5] The conclusion is objectionable if only because we recognize it as the ethic of an economic system that is bringing the world down around our ears. If this is the fruit of reason, then reason's axioms must be wrong. Something essential has been misinterpreted or omitted from Western thought.

The missing link is between people and the very world of Nature to which lip service is ritually paid but which nevertheless remains vague, unimportant and outside the pale of religious, philosophic and everyday thought. The profound thinkers of the West have cogitated much about their individual selves, about society and about

God, without scratching the surface of who they are, where they came from and where they are now, biologically and ecologically.

Religion and science have encouraged the arts to support the fiction that Nature is the alien other—hostile, untrustworthy, unlovable. The renaissance idea that opposed Nature to mind and rationality is still alive. The theologian C.S. Lewis, for example, stigmatized Nature as mere residue, after all of human worth has been wrung out of substance.[6] The conviction of the Manichees that the material universe is the devil's own kingdom of darkness against which the spark of spirit in humanity wages war, has survived close to the surface in the Judaeo-Christian tradition.

In order to buttress the idea of one transcendent God, the desert prophets and their successors denatured animate Nature, suppressing the reality by destroying the symbols: Baalim, Dionysus, Bacchus, Pan.[7] Having demythologized the world, having rid it of Pantheism, orthodoxy conferred on the transcendent God of justice and love a just and loving concern for humankind alone. Accordingly, Nature and her other creatures have been stamped as unimportant save as provisioners, no more than the means to support a humanity temporarily stranded on the waystation Earth, en route to a more important existence hereafter. The highest extra-personal ethic this value system yields is stewardship, whose goal of conservation is laudable as far as it goes but whose basic assumption—that people have the God-given right of dominion over all creation—perpetuates arrogant pride and the people/Nature rift.

Science, in turn, needed the missionaries of transcendence to prepare the way for its imperialistic onslaughts. Long regarded as the best kind of human knowledge, science has for four hundred years been an instrument for manipulating Nature rather than the means and encouragement for humanity to participate more knowingly in Nature's processes.[8] A powerful people-serving science could only develop in the Western world from which the Nature gods had been removed, pushed out by a historical God whose transcendent character also detached Him from Nature and thereby removed its spiritual values. Explanation, prediction, control—the aims of the knowledge-for-power syndrome—require a reductionist approach that anatomizes and dissects. The scalpel-wielding hands might tremble were scientists to entertain the belief that spirit is inherent in all things. That particular thought had to be banished in order for science to proceed.

"The mother-Earth concept," said a soil scientist recently, "although potentially interesting in examining man's relation to the land, is a hindrance to objective study of the soil."[9] If mother-Earth is to be skinned alive, her epidermis must first be given a value-free name such as "soil" or "dirt" or, best of all, "a mixture of chemical compounds."

Further, the tunnel vision of Darwinism, which revealed Nature essentially as species in conflict front-and-centre on the evolutionary stage, has foisted on all literate humanity a biased and ridiculous conceptual view in the name of science. "Nature red in tooth and claw" is apparent only to those who ignore the pervasive and underlying symbioses, co-operation and mutualism that necessarily constitute the environment of evolving life. The invention of a fiercely competitive Nature has justified competitive people and, in pursuit of the arts and sciences, has required egotistic personalities. But compliance, not competition, is what keeps complex life-sustaining systems on an even keel.[10] This is not romanticism; it is the reality of a sustaining world and of the human sub-world. Rather too slowly, the idea of Nature as an ecological system, or ecosystem, is redressing the excesses of prestigious Darwinian thought.

Ecology, like art, is a joiner, a perceiver of relationships, particularly within whole ecosystems. The living film at the Earth's surface—the animated Ecosphere bubble that encloses the planet's rock core—is the only complete ecosystem, encapsulating us and everything that is life-giving. The task today is to experience the in-ness, to feel it, to believe in Nature's beneficence, and thence to discover humanity's just place and humane role in the universe.

Moral philosophers, diagnosing the twentieth century, find humanity ill, perhaps at death's door. MacIntyre draws a parallel between our times and the epoch in which the Roman Empire declined into the Dark Ages:

> A crucial turning point in that earlier history occurred when men and women of good will turned aside from the task of shoring up the Roman imperium and ceased to identify the continuation of civility and moral community with the maintenance of that imperium. What they set themselves to achieve instead—often not recognizing fully what they were doing—was the construction of *new forms of community* within which the moral life could be sustained so that both morality and civility might survive the coming ages of barbarism and darkness.[11]

And so, remarking that the barbarians are not pounding on the gates but have been governing us for years, MacIntyre suggests a different kind of community to weather the approaching storm.

To a moral philosopher, "new forms of community" doubtless mean better human associations. That can be a start, though not imaginative enough. People as anxious selves banding together to save people is just not good enough anymore; the invitation is still to "get back to the party" with Bill. Unless the new forms of community extend beyond traditional humanistic bounds to include the community of Nature, the game is up. Art's new metaphor must be this larger universe.

The centuries-old separation of people from Nature took away from the world the values that consciousness finds in itself: intelligence, sensitivity, compassion, morality. When the self recognizes its source, these important attributes are returned to Nature, which becomes the bearer of values. If I as part of the world am alive, must not the world in some sense be alive? If the organic parts of ecosystems are intelligent, must not ecosystems, in some sense, be intelligent? Nature, no longer separate, is the key to fundamental revelations—the cosmic rhythms of dying and being reborn, the continuous passage from non-being to being, the paradigm of mythic time that rescues us from the terrors of historical time.

When the self recognizes its source, the sanctity of Nature is restored. In theological terms, the Deity is once again recognized as immanent, not just transcendent in far-off places but here, working in the world. Then the expression that Thomas Mann put in the mouth of his picaresque hero Felix Krull makes sense: "He who really loves the world, shapes himself to please it."[13]

To imitate Nature, to join her and be bound to her rather than seeking always to transform her, is the goal that could rescue the race from barbarism and darkness. To rediscover and strengthen our roots is not to regress but to grow.

If, within Nature, humans are conscious, ought we not consciously to strive to be the consciousness of Nature? And artists with a conscience, ought they not to point the direction, symbolically creating new realities whereby humanity may glimpse what it means to please Nature?

GROWING UP

GROWING UP IN GRANUM

When I was a kid growing up in Granum I never had my own gun, but that didn't stop me from hunting alone or with the gang. From Grade Eight on, somebody always had a Daisy BB gun or a bolt-action Remington .22 and after school or during the summer holidays, we spent a lot of time stalking the flocks of nervous sparrows that nested in derelict threshing machines at the edge of town, the pigeons that hung out around the grain elevators and especially the gophers whose appearance in March/April, amid lingering snow patches, heralded the authentic arrival of spring.

In those youthful days, the urge to hunt was strong, and the thrill of the chase intoxicating. Looking back, it seems that male members of my cohort spent an inordinate amount of time devising missiles of various kinds to extend their reach and to strike at a distance. The usual targets were animals large enough to attract attention, even our own kind on occasion.

In these violent pursuits, no adults coached us; we soaked up what was acceptable from the surrounding culture by osmosis, devising our primitive pastimes more or less naturally and spontaneously like Ernest Thompson Seton's little savages.

Ortega y Gasset, the patron philosopher of sportsmen, declared the chase dignifying and ennobling.[1] Glancing backward, down the long evolutionary road, he discerned Man the Hunter, senses alert to the sights, sounds, smells, the *feel* of nature, and he concluded with Paul Shepard that hunting is the generic way of being a Man.[2]

The hunt, according to Ortega, has both a mystical and ritualistic intent. Substituting a camera for a gun, a picture for a trophy, just will not do. One does not hunt to kill; one kills to authenticate the hunt, to elicit its true significance. Further, the sacrificial aspect is a source of virtue. Because man is a hunter, he is the most deeply loving and profoundly compassionate of animals: the Gentle Carnivore.

Were we youths unwitting participants in the Great Drama, practising to be charitable and merciful, learning to be truly human? Strange that in our councils, as in those of Ortega and Shepard, the virtues suspected of being feminine were excluded. Girls were expected to be camp followers, their ideal role applause from the sidelines.

Before guns came into our hands, we made bows and arrows, the bows from curved poplar branches, the arrows split from cedar shingles and whittled to shape. Sharpened at the heavy end, often inset with a shingle nail, the projectiles were designed to pierce and stick. One cold winter day, Orville Edwardh winged a finely crafted arrow clean as a whistle through two frozen sheets hung side by side on twin clotheslines, earning widespread admiration from all except his mother. And once, for a few days, our Buff Orpington hen carried a side-notched arrow around in her tail feathers, souvenir of a briefly popular atlatl style of dart-launcher.

We were also adept at making wooden guns sawed from the inch-thick ends of apple boxes. A trigger clothespin on the back of the pistol grip held—and with pressure, released—an inner-tube elastic stretched over the end of the "barrel." No threat to wildlife was posed by these short-range weapons, whose purpose was intra-specific combat. Their evolution was instructive.

True to military tradition, much effort was devoted to improving the power and efficiency of the guns. Initially an advantage in threat and deterrence was sought by broadening the gun handles to accommodate extra clothespins so that several rubber bands could be loaded. When the opposition caught up with the multiple shot design, the guns were made longer and longer until creeping technology produced the ultimate Big Boy: a hockey stick loaded with six knotted rubber bands stretched near the breaking point, a fearsome weapon that none could outface.

More lethal were slingshots fashioned from the crotched branches of Manitoba maple, with twin strips of inner tube tied to the arms of the Y and bound at their other ends to the squared-off leather tongue of an old boot. Such home-made catapults could hurl a symmetrical stone a hundred yards, and we searched the gravel roads for ammunition of just the right size and shape. Best were marbles, too precious for casual use on tin can and bottle targets, saved for close-in shots at whatever chanced to be the quarry of the day.

When Granum, like Troy, is dust, its fertile soils eroded, archaeologists will find among the scattered stone projectile-points of the earliest migrants, an equal number of "glassies" and the odd cat's-eye alley, mute reminders of a warring culture that also boasted heroic deeds.

For battles there were, sometimes spontaneous to avenge alleged insults but mostly staged and ritualistic. Through the shallow badland ravines of Willow Creek two miles west of town, we stalked each other in mock warfare, spying on troop movements and firing into enemy trenches. It was my fortune one summer day, to slingshot a stone that hit Dud Boyle between the eyes, laying him low for half an hour. By such mischances are reputations made and the accolade of "dead shot" was mine for a year.

Thorstein Veblen interpreted fighting and hunting as the twin marks of primitive, predatory civilizations.[3] Unlike Ortega, he used the past as a foil for the future. To the barbarian, he said, honourable is formidable. Arms are honourable, and the use of them, even in seeking the life of the meanest creatures of the fields, becomes an honorific employment.

The barbarian virtues are those designated manly: habits of ferocity, habitual bellicosity, histrionics expressed in the martial spirit and chauvinistic patriotism. Such virtues are displayed in the aggressive activities of boys. If persistent in the adult, said Veblen, they are the marks of an arrested spiritual development; the individual does not emerge from the puerile stage. The boyishness of most sporting men, including the exclusion of women from their activities, is notorious.

Were we Granum boys reliving humanity's childhood? Were we honing our incipient humanness, sharpening an already keen awareness of the natural world? Or were we passing through and recapitulating, without war-paint, an earlier stage, now indecorous, in the evolution of the race? The question is whether we evaluate ourselves by what we have been or by what we might be, the hunter that childhood revealed or the more pacific person of my aspirations, sons and daughters of *Pithecanthropus erectus* or compassionate sympathizers. Veblen urges us to look ahead, Ortega favours the backward glance.

Armed with real guns our attention necessarily swung to species outside the pale—that is, to all creatures excepting the one to whom the *Thou Shalt Not Kill* commandment intermittently applies. The accessibility of sparrows and pigeons as targets was somewhat reduced by their domestic association with buildings and glass windows around town, while out in the countryside, crows, magpies and hawks were too wary to be approached. Thus it was the misfortune of the peaceable seed lover, *Spermophilus richardsonii*—gopher, flickertail, picket post, Richardson Ground Squirrel—to bear the brunt of juvenile ferocity.

Gophers are fewer now, victims of mechanization that replaced the horse with the tractor, which was then used to plough and destroy the farm pasture, habitat of all native vegetarians and of other animals that depended on them. As the prairie pasturelands disappeared, so did the gophers, along with burrowing owls, weasels, badgers and ferruginous hawks, a careless and reprehensible trend that still continues. But back in the '20s and '30s, gophers were still the essence of the country landscape, contributing to the spirit of place along with flaring red sunsets and cactus flowers, wolf willow and crocuses, smells of sage and wild bergamot, sounds of cheerful meadow larks and plaintive curlews.

How well I knew the little fossorial rodent with its soft buffy underparts, its delicately mottled back and matching tail, the fine sheen of its fur. The head smooth, the eyes black and bright, the ears only minute apertures appropriate for an animal darting through tunnels. Locomotion: running on all fours but with an occasional clumsy comical two-legged leap. Voice: a high musical peep, delivered from an upright or crouched position, accompanied by a flick of the tail and a convulsive lift of the shoulders. Alarm signal: when chased into its burrow, a descending series of peeps, reverberating faintly in the Earth with a dying fall, suggesting a far away retreat into mysterious inaccessible subterranean corridors. Prolific, producing playful inquisitive innocent young, suckers for the snare and the trap.

And so we snared them, trapped them, stoned them, drowned them out and gunned them down. Too unsophisticated to use the language of sportsmen, we did not "harvest" the little beasts, we simply killed them.

A .22 bullet slamming into a gopher makes a dull thud. "I got one!" the hunter shouts when, on pulling the trigger, he hears a *plop* rather than a *whang*. Then a rush to the burrow to find yet another bloody gopher expiring, convulsively thrashing and kicking its life away.

One day when I was thirteen, a family friend came by to take my father and me for a visit to his farm. In his truck he carried a .22 rifle and, as a generous host, he stopped en route so that by way of recreation we could shoot a few gophers; "They eat the crops." I had never been out on a hunt with Dad, vaguely sensing that he didn't share my enthusiasm for the chase. Nevertheless when it came his turn to fire the gun, I was astonished that he could not hit the gophers that stood up chirping at him from twenty yards away. Three times he shot, the bullets harmlessly kicking up dust in the distance.

Then an improbable thought forced its way into my consciousness. Dad was purposely missing! I could read it in his face, in his unhappy eyes. He was trying *not* to kill the gophers! Mild embarrassment tinged my surprise at this breach of convention. But three years later one winter day, stealthily hunting in the willow bush along the Highwood River, I shot a snowshoe rabbit. It screamed when the bullet struck. I looked at its lustrous eyes and its soft white fur and remembered my father's marksmanship that suddenly seemed far better than mine.

That day was my last as a hunter. I committed myself to Veblen's camp. The Gentle Carnivore set aside his rifle—but went home to a chicken dinner.

BEAUTY AND THE BOTANIST

Flowers stir our sense of beauty and lend a special charm to botanizing, whether in our gardens caring for roses and sweet peas or rambling in nature's paradise searching for rarer blooms. Flowering plants evolved along with our mammal ancestors; we grew up together during the last few hundred million years, and the roots of the attraction we feel for fragrant blossoms and bouquets lie deep within us.

Professional botanists shy away from the subject of beauty in the plants they study. Such neglect by science's practitioners deserves attention in a fading world where the common un-beautiful arguments for preserving wild landscapes, and the plants that are parts of them, continually fail.

We have heard again and again the reasons for preservation of at least a few viable parts of the non-humanized landscape, reasons that always emphasize benefits to the human race. The arguments rely on the logic of economic utility, stressing usefulness to people, to society, to science. Their weakness lies precisely here, for everyone, with a little thought, perceives that the utility of wild things as such—of landscapes in their original forms and of the native plants and animals within them—is small compared to the utility that interventions and developments will yield in material "goods." Strict protection, hands off the natural world, butters no parsnips.

Better to farm and lumber, mine and hunt over all of Canada. Unless of course the land is relatively valueless for agriculture, forestry, mining and hunting. Then society concedes that, yes, perhaps a park *is* in order—to draw in tourists and give a boost to the local economy. The promise of profitability-through-use seems today the only acceptable excuse for salvation of the world. But acceptance of that rationale opens the door to judging "best use" by market values, against which the "let it be" philosophy cannot compete.

The economic justification is weak because it is wrong. The battle for preservation will be won, if it can be won, in the arena of non-utilitarian values, with no price-tag attached. Only immaterial values—appreciation of landscape beauty, the sense of intrinsic values in Nature—will finally serve.

The time has come to champion the aesthetic sense and trust it, rather than relying on the economic sense that calculates short-term market profits and discounts the future. Enough of crass assertions that humanity can profit from protected areas because eventually they "payoff": as benchmark sites by which we can gauge the extent of the world's deterioration; as models for reconstituting ecosystems we have ruined; as gene pools from which to engineer productive organisms; as sources of still-undiscovered crops, domestic animals, pharmaceuticals. These shabby arguments will not preserve the natural world—although they may successfully preserve the public funding of research. The alternative is faith in the affections that gently lead us on to a sympathy reaching out beyond the human race.

UNAFFECTIONATE BIOLOGISTS

Professional biologists make no concessions to beauty. By a conscious effort they exclude from both subject matter and personal attitude all considerations of beauty and such soft-hearted affective values. The reasons are hard-headed. *Control* of phenomena, not *sensitivity* to them, is the goal of science. Scientists search for mechanisms within, not aesthetic emanations without. Quantitative measurement is the written language, and neutral numbers and formulae—no matter how elegant—are incapable of expressing beauty. Aesthetic feelings, if countenanced, would distract from the accepted analytic methodology. Further, scientists are cut-ups, and dissection is their proper tool. To ask if the living whole might not be more beautiful than the dead parts would subvert the entire enterprise.

The danger of the scientist's belief system is its mechanistic materialism, its goal of control and its dispassionate stand-apart-from-the-world methods. They erode and eventually eradicate those non-utilitarian instincts that we must rely on and trust if the non-humanized natural world is to be preserved. The peril has long been recognized by the radical fringe, and the poetry of Blake expresses it with profound insight:

> To see a World in a grain of sand,
> And a Heaven in a wild flower,
> Hold Infinity in the palm of your hand,
> And Eternity in an hour . . .
> A skylark wounded in the wing,
> A cherubim does cease to sing.
> The wild deer wandering here and there,
> Keeps the human soul from care.[1]

How dull on purpose are the abstracts of the papers from the annual meetings of the Canadian Botanical Association! No one discusses "The Beauty of the Flower," "The Esthetics of Riparian Vegetation" or "Sensory Delight in the Study of Organelles." Smart young botanists soon discern the range of legitimate interests dictated by a professional career: acceptability by the established peer group and the necessity of publishing in reputable journals. They learn the lesson that science's content excludes beauty and has naught to do with qualitative aspects, with emotions and feelings. Such fuzzy aspects of experience are not merely banished to the realms of art and culture; they are obliterated.

SCIENCE EXCLUDES QUALITY

The rational mind of science has a terrible failing. If it cannot cope with aspects of experience that are un-measurable, it declares them meaningless, unimportant. They are unsubstantial epiphenomena. Feelings, the promptings of emotions, are distrusted because no white-coated savant has corked the incontrovertible evidence of them in a test tube. The better humanity becomes at science and its kind of rationality, the less room remains for non-cognitive things of the spirit. Values still call society's shots, but they are debased to those values that can live with science and those with which science can live. Truth is one such precept, but truth-in-science is a peculiarly denatured form of the real thing. The history of science explains this incongruity.

Galileo is one of the heroes of rational thought, and the story of his persecution by the Church is legendary. Time has acquitted him of the Inquisition's charge that he blasphemed in rejecting the idea that the solar system revolves around Earth. But Galileo did commit a serious crime: the rejection of total human experience. He divided experienced reality into an objective quantifiable sphere, the real world of Primary Qualities and science, and a subjective qualitative sphere of feeling and meaning, comprising unreal and unimportant Secondary Qualities:

> But I hold that there exists nothing in external bodies for exciting in us tastes, odours and sounds but size, shape, quantity and slow or swift motion. And I conclude that if the ears, tongue and nose were removed, shape, quantity and motion would remain but there would be no odours, tastes or sounds, which apart from living creatures I believe to be mere words.[2]

All but shape, quantity and motion are "mere words." Galileo elevated sight and touch as the senses that connect us to reality, thereby dismissing the more emotive mind, the aesthetic mind.

From Galileo's time to the present, the world's physical scientists—acclaimed as the weighty thinkers—have insisted that Nature speaks only the language of quantity, of mathematics, in a mechanistic way. Nature is not organic but mechanical. In this century, the philosopher-mathematician Alfred North Whitehead objected to the non-organic world view, characterizing materialistic scientific thought as "one-eyed reason, deficient in its vision of depth." Nevertheless, despite Whitehead, and before him those in the tradition of Rousseau and Thoreau, scientific materialism has dominated and spread around the world. In the twentieth century, the majority of people value power and control over the material universe above all else, and science delivers that power and control in spades.

This is not to attack science in all its human knowledge forms. As conscious beings, we *will* have knowledge, inescapably, of one kind or another. But criticism of a certain kind of power-seeking knowledge—science as commonly understood—is merited, as is the widespread belief that the pursuit of this control-oriented rationality is the most important human endeavour, to which all others—including search for the good and the beautiful—are subordinate. Fortunately intelligence is not so confined.

SCIENCE FILLS THE FAITH GAP

Journals regularly run articles decrying the few numbers of women in science, particularly in the physical sciences. By implication, women are the halt and the lame who need to be healed and made whole by better education that will turn them into what they ought to be: researchers and academics. True, women have suffered from inequality in our educational system; they have been denied access to various fields and have been short-changed by sexual stereotyping. Nevertheless, to fasten on the unjust socializing process as the sole reason for relatively few women in science is as unwise as to blame school curricula for not making all men mathematicians, physicists and engineers—a goal that the Science Council of Canada would doubtless approve.

I suspect that many women, and a large number of men too, get bad vibrations from science. Substantial numbers of both sexes do not pursue science careers because in a deep and intuitive way they sense that dispassionate intellect, quantitative rationality, leave out too much of what is real, vital and believable in their lives. They sense that science is not the solution to the human predicament but a problematical part of it.

Each of us has to believe in something that helps make sense of the world, for the absence of ways to rationalize events, to impose order on disorder, to create a world of meaning, is the road to madness. Caring for things beyond our immediate selves is another requirement of sanity. Organized religion used to fill these dual needs, explaining life's purpose and providing an outlet for unselfishness. Now, for many, the old beliefs have faded. Astronomers searched the heavens and found them dead and mindless; Kruschev reported that USSR astronauts had spotted no God in outer space. Many people were left with nothing to believe in, nothing worthy of worship but their own kind. Species selfishness was enthroned. The popular religious sects preach a Dale Carnegie brand of success through personal salvation. Secular society is pragmatic, concerned with the economic and social welfare of its members.

The vacuum left when old-time religion was discarded has been filled without fanfare by the religion of Humanism whose goal is service to humanity and nothing else. Today all right-thinking people and all right-acting institutions make helpfulness and service their motto. Notice all the selfless assistance offered by the banks and other corporations, the fraternal clubs, the insurance agents, your friendly corner service station. When service to people is paramount, disservice to everything else in the world follows close behind. How ridiculous the suggestion that we should stop using animals in research! Without cutting them up and dosing them with chemicals, how would we ever serve people medically, how find cures for their diseases?

The accent on service has been co-opted by commercialism and greed. The world of advertising makes a mockery of beauty and altruism and sacrifice. Reading the ads turns the sensitive cynical and sick. What is left? Where can idealistic young people turn if they reject the hypocrisy of the business world and a merely personal salvation?

Science seems to offer a worthwhile escape, for it holds up a selfless banner promising truth, progress, freedom, a better world. Science in fact has become Humanism's chief religious sect, not lacking in dogmas and Latin maxims. The motto of my Alma Mater—*Quaecumque Vera*—"Whatsoever Things are True," has a nice ring. But now we are told by such philosophers as Kuhn that "truth" in science means correspondence with the latest theory, with the paradigm of the month. Such truth has little relevance to the preservation of a habitable world or to creating a decent society within it. Science's truth is not the "Beauty is Truth, Truth Beauty" of the poet.

TEMPLES OF SCIENCE

In the golden age of universities, early in the century, Einstein wrote these words honouring Max Planck:

> In the temple of science are many mansions, and various indeed are they that dwell therein and the motives that have led them thither. Many take to science out of a joyful sense of superior intellectual power; science is their own special sport to which they look for vivid experience and the satisfaction of ambition; many others are to be found in the temple who have offered the products of their brains on this altar for purely utilitarian purposes. Were an angel of the Lord to come and drive all the people belonging to these two categories out of the temple, the assemblage would be seriously depleted, but there would still be some men, of both present and past times, left inside. Our Planck is one of them, and that is why we love him.[3]

Einstein's imagery of temple, altar and angel places science in a religious context that the gothic architecture of many university campuses reflects. Humanity progresses under the leadership of an unselfish, dedicated priestly elite. The direction is unquestioned. Twenty three years later, however, in 1941, a more pessimistic but clear-sighted Einstein wrote of science:

> Whatever this tool in the hand of man will produce depends entirely on the nature of the goals alive in this mankind. Once these goals exist, the scientific method

> furnishes means to realize them. Yet *it* cannot furnish the very goals . . . . Perfection of means and confusion of goals seem—in my opinion—to characterize our age.[4]

Einstein came to recognize that science is instrumental, a tool, a means. As long as we are sure where we want to go with science, it is wonderful. Science will take us in whatever direction we choose, toward worthy or unworthy goals with equal facility.

From this proposition a corollary truth: science left to itself and setting its own expedient directions will do as much evil as good. Nevertheless, every professional worth her/his salt argues that s/he should yearly be given a bundle of money by the National Science and Engineering Research Council and then be left alone to do her/his thing. This is the dream of all scientists, endorsed by prestigious Nobel Laureates, that society regularly and unquestioningly provide ample no-strings funding for the pursuit of "basic" or "curiosity-oriented" research. Be assured, says the National Research Council, that nothing but good will result.

But hold on a minute. Where did brainwashing, thought-control, acid rain, ozone holes and nuclear weapons come from? Ah, the scientists reply, you are failing to distinguish between our spotless, dispassionate, objective science and the unfortunate and messy technological uses to which it is sometimes put. Do not blame us if the pure and neutral knowledge we deliver is sometimes misused.

That answer is at odds with Einstein's conclusion. In none of its forms is science isolated from society. It is a means, tightly tied to the technology it spawns, and both are power tools serving human goals. Science cannot escape the blame when the potency it releases is used destructively.

A NEW GOAL FOR SCIENCE

A vital question, then, concerns the ends that science/technology serves: the question of goals and guiding values. Although these can be examined rationally, the intellect is not their source. The leading goals for society derive much more from sensibilities—feelings of harmony, beauty, sympathy and justice—than from pure intellect. The temples mistakenly erected to science on university campuses should be rededicated to value-oriented wisdom, to a new faith that transcends instrumental knowledge.

An attractive belief in the importance of Nature is resurfacing. Tribal cultures and those of our ancestors have conceived humanity and the world as organically one. Today, ecological insights are buttressing the credibility of this ancient faith. A recent restatement is the Gaia Hypothesis by James Lovelock, identifying the Earth as a vital being of which all organisms along with the air, water, soils and sediments are dependent parts.[5] Gaia in Greek mythology is the Earth Goddess, and humanity is likened to her son Antaeus who retained his strength as long as he was in touch with his mother. Whatever separates humanity from its Earth roots—present-day Herculean technology, for example—saps health and wholeness. (In the myth, Hercules overcame Antaeus by holding him aloft, so that he could not touch the Earth.) Enlightened science/technology will strengthen the bonds between people and the Earth, not destroy them.

To conceive the Earth as the creative centre requires another Copernican revolution, a new paradigm shift that removes the focus from people and places it on their Home Place, the planetary Ecosphere. A greater than human reality would then revolve around the Earth, recognized as the source of creativity, life and health. An expanded religious sense, recognizing divinity in surrounding nature again, would relieve our inturned, suffocating and stultifying preoccupation with The One Important Species.

With such a faith, how optimistic the future will be; our twin science and technology under control, both the twins guided in a healthy direction as humanity harnesses its intellectual powers to rescue the planet, to rehabilitate and maintain its beauty, permanence and productivity. Reconciliation with the Earth will expose the foolishness of death-wish military technology that currently wastes a trillion dollars a year and destructively employs half the world's scientists.

The beauty perceived in nature is the fragrant aura of the old/new Gaia myth. The perception of beauty in the world is the proof of our belonging, the bridge to a wider sympathy extended beyond our own kind. Beauty is not only what we perceive to be harmonious and healthy; it is a central aspect of whatever we love.[6]

Imagine how a softer science, infused with affection for the world and its marvellous parts, could contribute to compassion for all things, both living and supportive of life. Why not a symbiotic science of botany that marvels at the beauty of flowering plants and explores their many meanings? Botanists of the world, wake up and smell the flowers.

THE LAKE ATHABASCA SAND DUNES

Marvellous is the word for the great sand dunes of northern Saskatchewan and adjacent Alberta. Viewed in the dazzling light of long summer days they seem an enchanted landscape—a portion of Arabia magically transported to the boreal forest and displayed along the south shore of blue Lake Athabasca.

The illusion is enhanced for those interested in natural history, because this is the home of unusual landforms and unusual plants. Inland from the lake's beaches a sandy plain rises, divided by streams, lakes and park-like strips of spruce and pine into seven major dune-desert areas and a larger number of small ones, comprising, along with gravel plains, scoured rocky knolls and old strand lines, the bed of a more ancient lake. The vegetation is sparse but intriguing, with about sixty plant species designated as rare.

The spectacular shifting sands constitute the largest active dune environment in Canada, as well as one of the largest in the world north of fifty-nine degrees latitude. Spread over a belt of some 1,500 square kilometres, largely between the major north-flowing rivers—MacFarlane River on the east, Dumville Creek and William River on the west—the extent of the dunes is such that from a distance, on cloudy days, they project on the sky a strange coppery glow. Small wonder that the native people have regarded the terrain with awe, recognizing its unique spirit-of-place in a region otherwise dominated by coniferous forest and peatland.

The open dunes occupy the northern edge of the great Athabasca Sandstone basin—horizontally bedded rock 1,500 metres deep at the centre, thinning to the edges where uranium ore bodies lie shallowly buried. Oval in shape, the basin's longer axis extends east-west from Wollaston Lake into northeastern Alberta. Cree Lake lies at the south end of its shorter axis.

In Precambrian times, the basin was a large delta receiving sand

and gravel from eroding highlands to the east and southeast. These same sediments—weathered again into sand and exposed to wind action by glaciation and the post-glacial falling of lake levels—are the source of the dunes. Streams carry the sand northward to deltas from which currents spread it along the lake shore. When lake levels drop, as they have intermittently ever since the last glaciation, broad beaches are exposed to the prevailing northwest winds that recycle the sand inland again.

Isolation has helped to maintain the pristine beauty of the landscape. The nearest settlement is the ghost town, Uranium City, forty kilometres across the water on the north shore. The Cluff Lake uranium mine and mill, with a short life expectancy, lie sixty-five kilometres south of the shoreline, joined to the lake by only a winter road. Apart from a declining fishery, no money-making resources sufficient to attract roads and vehicles have been identified. Access is still by boat or float plane, and the exposed shoreline—unpredictably battered by high waves—makes landings and departures hazardous. Visitors must be prepared to stay when the weather changes for the worse and strong winds blow.

The south shore was visited by Harvard botanists Hugh and Lucy Raup in 1935. The wife and husband team made the as-yet-unexplained discovery that the dunes and gravel plains harbour a small group of plants found nowhere else in the world. Of the ten most rare, four are willows and six herbs. Silvery leaves are characteristic of four of the group, an adaptation not uncommon in desert environments. This botanical find was startling because endemic plants are practically unknown in the rest of Saskatchewan and adjacent Alberta, nor are they usual in the Mackenzie River drainage basin to which Lake Athabasca is joined.

The last glacier cleared the northern boundary of Saskatchewan at sixty degrees latitude only eight thousand years ago. Did the plants evolve since then under the intense selection pressure of the rigorous environment? Or did they evolve elsewhere and immigrate, finding here their final sanctuary? The puzzle awaits detailed study, along with questions concerning the distributions of the other fifty rarities. Sea-lyme grass, for example, a food plant of the Vikings, grows profusely along the south shore. How did it happen to arrive? Did human migrants bring it in?

The terrain must always have been attractive to wanderers. Early aboriginal visitors left records of themselves in sixty archaeological sites so far discovered. Palaeo-Indians camped along the old strand lines between eight thousand and seven thousand years ago when the Lake stood almost one hundred metres higher than today. Later arrivals, identified by different stone tools and projectile points, represent cultures of the southern Shield Archaic and ancestral Inuit of the northern Arctic Small Tool tradition.

The most recent artifacts indicate the meeting of both western and eastern native groups, doubtless bringing together technologies from the Plains and from the Boreal Forest. Today an area bordering the MacFarlane River has been claimed as traditional land by the Chipewyan of Fond-du-Lac.

One of the first Europeans to pass through was Philip Turnor in 1791. He left little account of his visit, perhaps disappointed that Fort Chipewyan at the west end of Lake Athabasca was not near the Pacific Ocean as those who sent him out from England had hoped. Of the dunes west of the MacFarlane River he wrote: "Sandy hills drifted like snow drifts in ridges . . . heigh sandy hills which when seen off the mouth of this river looks like fields of ripe corn between ledges of woods such as are seen in the Hill countrys of England."[1]

Turnor's journal entry also noted the legend of the "Chepawyans" Giant Beaver, "which formerly was in this part and which turned up all this sand but their was a species of Giant Indians who killed all those Beaver and whose race is likewise extinct."

A century later the geologist-surveyors from Ottawa arrived: R. G. McConnell who named the Athabasca Sandstone in 1888 and D. B. Dowling who travelled up the William River in 1892 and wrote, "On the surface of the plateau which is mostly bare, sand-hills rise in some cases nearly a hundred feet above the general level."[2] Despite the geological interest of the area, another eighty-five years elapsed before Brian Schreiner of the Saskatchewan Research Council mapped and described the surficial geology as part of a cooperative study with the University of Saskatchewan.[3] An important analysis of the terrain and vegetation was contributed at about the same time by Hugh Raup and George Argus.[4]

Aerial photos disclose the variety of fascinating landforms in this apparently misplaced "desert." Eolian forms, those shaped by wind, include parabolic dunes in hairpin and boomerang shapes, often overlapping in embroidery or rope-ladder patterns or with long ridges trailing upwind from the extremities of their arms. Others west of the William River are high, elongated, sharp-crested single ridges, resembling the sword-like seif dunes of the Sahara.

Wind directions of the past are preserved in dunes that point northwest, showing a shift of 180 degrees from today's prevailing gales. Some of them, formed close to the ice-sheet and more than eight thousand years old, are stranded in the middle of peatlands, an indication that the climate was much drier when they were active.

On level areas flanking the dune systems, the wind has winnowed away the fine materials, concentrating stones at the surface. Sandblasted over the centuries, many have been cut to the shape of Brazil nuts: three-sided polished stones called ventifacts or, in German, "dreikanter." The stony plains are "gobis" or desert pavements that shine in the sun. Only one stone deep over soft sand, they are sensitive even to the traffic of bears and moose. All-terrain vehicles would quickly destroy them.

No less interesting are features within and at the edges of the dune fields: streamlined hills and flutings sculpted by advancing ice; a series of remarkable little single-ridge end-moraines that show where the last ice-sheet paused temporarily at least six times in its northeastern retreat; meltwater gravel ridges that enclose gem-like kettle lakes.

Bounding it all on the north side, outstanding in its own right, is the shoreline of the lake, forty miles of broad beaches looped in gentle crescents between the headlands—William River delta, Beaver Point, Turnor Point, Wolverine Point, MacFarlane River delta. A saunterer's delight, the strand offers a clean firm trail to be explored and savoured, mile after mile.

For this is great hiking country, whether by the lake shore, following game-trails along the streams, under the open pine woodlands with their attractive lichen ground cover, or back in the quiet solitude of the brooding sandhills. Those who have been fortunate enough to camp and live there for awhile know the exhilaration of the experience. And yet the very attractiveness of the area and the obvious fragility of some of its parts raise disquieting questions about our attitudes toward rare wildernesses and our presence in them. Should we think of them as sacred places? Much can be re-learned from the indigenous people whose languages contain no equivalent of "wilderness."[5] Feeling themselves part of the Earth and not owners of it, they had no need for words that separate people from nature, the domesticated from the untamed, the city from the wilderness. We who came with ideas of dominance and control backed by powerful technology are beginning dimly to realize the need to regain what the native people intuitively knew. If we are to progress together, the erstwhile know-it-all teacher must become the student.

Wild areas preserved just for their wildness, simply for their non-humanized features, symbolize an emerging attitude whose other expressions are swelling interests in conservation, in sustainability, in combatting pollution, in saving other species from extinction, in stemming the destruction of tropical forests and the oceans. Countries that count themselves civilized need the symbolism of wild places to remind them of their sources and, in a deep way, of their continued co-existence in the world.

All who have seen the Lake Athabasca Sand Dunes water-and-landscape at first hand or in pictures agree that it is a unique treasure worthy of preservation, not only for the historical, cultural and scientific stories it offers but also for its inherent mystery and beauty, for its spirit-of-place. It has been named an area of exceptional significance by Parks Canada, and Park Reserve status for the sector on the Saskatchewan side was confirmed by the provincial government in 1989. Within five years, by 1994, decisions about the future of the tract and how we shall relate to it must be made. The opportunity to create an inter-provincial preserve, a dedicated wilderness that includes the Alberta dunes, should not be missed.

ARKS CAN'T SAVE AARDVARKS

Humans discovered their environment as all-encompassing with the first outer-space photos that revealed the Ecosphere: a blue cloud-swathed globe in whose watery skin a marvellous variety of protoplasmic bits and pieces are enclosed and sustained. Suddenly we were given a breathtaking new perspective on ourselves: self-conscious deep-air animals, a dependent and integral part of it all.

Suppose that this vision, this reality, had preceded the development of today's science. Suppose that we had been given the outside perspective to see the Earth entire, ourselves immersed in it, and had taken the vision to heart before we decided what was important, what was real.

Had such transcendent insight been granted, would we not have recognized the Ecosphere, the Home-sphere, as the unitary thing to be valued above all else? Only afterwards would we have studied and analyzed it, dissecting it into its component water bodies, atmosphere, continental platforms, plants and animals in their communities and as individuals—in order to better understand its miraculous integrity.

Unfortunately, submerged in it we were unaware of the whole. Like the blind men in the fable who mistook the legs or tusks, the trunk or tail for the elephant itself, we have assumed that the parts of the Earth are autonomous things-in-themselves, starting with humans and working out from there, identifying as most important the objects with properties similar to our own: animals and plants.

Various other things around us—apparently peripheral odds and ends—in turn forced their importance on our attention: the aerial climate, soils, sediments, salt water, surface and subsurface geological rocks and ores. These we called "raw materials" and "resources" when we perceived human uses for them, while to those dimly perceived to be indispensible we gave the name "environment," meaning that which surrounds objects of greater importance: namely, organic things like us. An inadequate conception, this view from inside.

The view from outside came too late. The un-sensed Ecosphere had already been dissected from the inside out. Philosophers and theologians had pronounced their final Truths about its parts. Disciplines had been defined and disciplinary knowledge had hardened. Universities and governments were de*part*mentalized to deal with the fragments, their adherents assured of certain certainties. In millions of books and learned treatises, the components of the Ecosphere were confirmed by the savants to be self-standing entities. "The proper study of mankind is man," we have been taught. "Soils are natural bodies at the surface of the Earth." "Rare and endangered plants and animals can be saved." All part truths and all misleading.

Ideas are changing. The new perspective from outside casts yesterday's assumptions in a critical light. More aware of our evolutionary history and our dependencies, we know intellectually that the segments of the world we study in the disciplines are indeed segments. Geoscience tells us that what we call atmosphere, lithosphere, hydrosphere and biosphere developed together; they have no separate reality, except in thought. Ecology tells us that what we call organisms and environment are inseparable, except as words. But these facts have not been assimilated. We do not know them in the way that counts: in our affections as well as in our heads.

How difficult it is to accept in our hearts and imaginations that the authentic thing, the important thing, the primary reality, is the Ecosphere, one of whose attributes is the phenomenon called life. Life is not a property of organisms, nor of cells, nor of strands of protoplasm, nor of complex molecules like DNA. Life is a property of the skin of the planet and of the ecological systems that the marvellous skin comprises.

One-eyed biology, lacking depth perception, has mislead us into conceiving a world divided into the biotic and the abiotic, the organic and the inorganic, the animate and the inanimate, the living and the dead. The divisions are false and mischievous. For what would qualify as animate, living, organic and biotic without sunlight, water, soil, air?

These latter components are as vital and as important as the organisms whose life-giving environment they are. The Ecosphere, the most perfect ecological system, comprises all-of-them-together. It is an evolving, adjusting, self-repairing *Supra-organismic Unit,* transcending the organic, and not just a super organism.

One of the tools of human understanding is reduction, anatomizing objects of interest into their parts. Into what parts should the Ecosphere be dissected to aid comprehension? The answer, the wrong one, has already been provided: primarily important are organisms, and secondarily their nebulous environments or habitats. But that is the myopic view from inside the system not the comprehensive view from outside.

If we take to heart the truths that the three-dimensional skin of the planet, the Ecosphere, is the unit of importance and that life is not a phenomenon that exists apart from it, then clearly the most significant parts are volumetric landscapes and waterscapes. Anatomizing the Ecosphere into sectoral three-dimensional ecosystems whose components include plants and animals along with their matrix of atmosphere, soil and water, provides simplified but almost complete homologues of the Real Thing. Such ecosystems, chunks of the Ecosphere, can very nearly exist on their own, like large terraria and aquaria. Plants, animals and people, individually or in their populations and communities, cannot do that!

Endangered organisms *per se* cannot be preserved. Ecosystems of which organisms are interesting ingredients can, however, be preserved—as long as the Ecosphere of which they are parts continues to function in the old natural and healthy way.

This realization should turn attention more and more to the absolute necessity of preserving wildernesses and natural areas, ecological reserves and sanctuaries, endangered *spaces* before endangered *species*. Unless natural ecological systems are preserved, the native flora and fauna will not be preserved. Organisms will still exist in the truncated environments fashioned by well-meaning people but only as cultivars, zoo freaks, the living dead.

In practical terms this means that we never deal just with organisms but always and necessarily with ecological systems of which organisms are notable parts. Preserving a rare plant population in any but a temporary sense must mean preserving the rare ecosystem of which it is one component among others of equal importance. Taking a threatened species on board an ark to save it from the industrial deluge is to immerse it in a tame ecosystem from which—if kept there for more than a generation—it will emerge as something else, on its way to becoming a tame species, a cultivar.

Unless our various arks—zoos, arboreta, seed banks and propagation gardens—are looked on as temporary lifeboats whose occupants are to be landed as speedily as possible on whatever Mount Ararats and other sanctuaries protrude above the human flood, all the current discourse about preservation will be only fine talk. The survival of some threads of biological history that we knew to be valuable may be secured, but their texture will no longer be that which drew us to them in the first place. Somewhere along the way, full of good intentions, we will have killed the things we love.

In this there lies a lesson for ourselves. We too have built and boarded an ark, a cultural vehicle that is carrying us into the future. We have cast off from our native shore, left it behind and plan never to land there again. Unlike Noah we have made room for few other passengers, preferring to exterminate rather than to save. Our brave new world will be a people-only world, an ark expanded—electronically and chemically bigger and better, like the metal-and-glass spaceships of science fiction—in time entirely replacing the old world. True, we have not thought out whether the idea is good or bad, but meanwhile build and sail onward, to a richer and more abundant future!

Humanity is already suffering from having been on board its particular monospecific ark too long, from solitary confinement in the plastic and asphalt environment, shut off by choice from its roots in the ancestral world, not enough bird song, too little spongy turf underfoot. More and more the human species is out of touch with Nature, too citified, at risk from its own seed-bank environment—rows and rows of boxes in which we have voluntarily imprisoned ourselves and where with less and less success we attempt to bring up good and natural progeny.

We will not save the riverine forests without protecting the floodplains, nor will the orchids be preserved without preserving the marshes. Our own fate is linked to the limits we set on the domestication of the world around us and to the offsetting effort we devote to maintaining the life-blood of the Home Place, the natural beauty and health of the creative, sustaining, enveloping Ecosphere.

STRICTLY ACADEMIC

CRIMES AGAINST THE ECOSPHERE

INTRODUCTION

The Law is the product of an evolving political process, reflecting issues deemed important by society and contributing to the choice of appropriate actions. Both issues and actions, political and legal, represent a selection among alternatives, with current ideology the screening mechanism. Political goals change as belief systems change, and, as part of the change (though lagging in the rear), laws are recast to conform to new realities. Environment is one such new reality.

The Law Reform Commission of Canada (LRCC) reports to Parliament through the Minister of Justice and was established in 1971 to modernize and improve both concepts and approaches to federal law. Its limited success in influencing Parliamentary legislation belies its educational influence. Recognizing that today's rapid social changes demand more than statutory tinkering, the LRCC has sought to clarify basic principles and has inquired philosophically into the values that the law should uphold. By this route it has come face to face with environmental problems and has produced several study papers and working papers in its "Protection of Life Series."

One of these, *Crimes Against The Environment*[1] is an innovative study. It heads into unfamiliar territory by recommending the addition to the Criminal Code of a new and distinct offence whose goal is the repudiation and deterrence of conduct "that seriously compromises a fundamental societal value and right, that of a safe environment or the right to a reasonable level of environmental quality."

Regardless of the success of this law-reform proposal, the subject is topical and more important every day. By what legal means is environment to be protected most effectively? Are the powers of the Code, just as it stands, sufficient for the deterrence J and punishment of those who actively or by negligence cause I significant environmental harm? Is recognition of a "new crime" against the environment necessary?

These are questions that lawyers can best answer once they clearly understand the object of concern. Most treatises that deal with "the environment" (or sometimes "the ecology") accept it as a given, as common knowledge, requiring no close scrutiny. The concept environment is assumed to be simple and uncomplicated, thoroughly understood by everyone—lawyers, philosophers, laymen. Unfortunately, the assumption is wrong.

The justification for my examination of the Working Paper is not legal but ecological, probing the premises on which the arguments for environmental protection, via the Criminal Code, are built. From a better understanding of environment, both the motivation and the means for its protection should emerge.

The stage is set by deciphering the Commission's rationale. Briefly, *Crimes Against the Environment* accepts the traditional anthropocentric (homocentric) position that takes *environment* to be exactly what its etymology suggests: the context and surroundings of things of greater importance—namely, people. In this popular sense, environment is peripheral; the concept is its own pejorative. Logically then, environment's defence must be restricted to terms of human utility; it is a "societal value and right," not a thing of inherent worth.

I argue that the alternative—recognition of environment's intrinsic values and thereby its inherent rights—provides the only incontrovertible basis for protecting it against crimes of despoliation and degradation. But first the misleading concept *environment,* as that which merely surrounds humanity, must be replaced with the concept *Ecosphere*—the creative evolved and evolving shell of air-water-Earth-organisms that mantles the world. The Ecosphere—the planetary skin—envelops, sustains and reproduces all organisms, thereby constituting, in the most fundamental sense, life's irreplaceable environment.

THE CRIMINAL CODE AND ENVIRONMENTAL PROTECTION

The published version of *Crimes Against The Environment* represents the fourth draft, doubtless reflecting the litigious nature of the subject as explained on the first page:

> We cannot claim that there was unanimous endorsement of our proposal to add a new offence of a 'crime against the environment' to the Criminal Code, but there is wide support for doing so . . . . As for particular aspects of the proposals, many judges and others fully endorsed the advisability of a generally worded formulation for the new offence . . . and agreed that the flagrant and dramatic violation of federal and provincial environmental statutes and emission standards should be a necessary condition for criminal liability under the proposed new *Code* offence.
>
> The present Criminal Code in effect prohibits offenses against persons and property. It does not, in any explicit way or direct manner, prohibit offenses against the natural environment itself. In this Working Paper the Commission makes and supports the proposition that the natural environment should now become an interest explicitly protectable in some cases in the Criminal Code. Some acts or omissions seriously harmful or endangering to the environment should, if they meet the various tests of a real crime, be characterized and prohibited for what they really are in the first instance, crimes against the environment.

When is pollution[2] a real crime? The Commission refers to its third report, *Our Criminal Law*,[3] in which five characteristics or signs of Code crimes are proposed. Offenses should be considered real crimes only if: 1) they contravene a fundamental value; 2) they are seriously harmful; 3) they are committed with the required intentional mental element *(mens rea)*; 4) the needed enforcement measures would not themselves contravene a fundamental value; 5) treating them as crimes would make a significant contribution to dealing with the harms and risks they create.

Applying these tests, the Commission concludes that certain

instances of "intolerable," "gross," "catastrophic" pollution qualify as real crimes. Just as some offences ought to be removed from the Code because they no longer are perceived as serious threats to society's fundamental values, so offenses that today are perceived to contravene nascent fundamental values should be added to the Code.

Therefore, in response to its mandate to develop new approaches to and concepts of the law in keeping with and responsive to the changing needs of Canadian society, the LRCC has concluded that certain serious instances of environmental pollution should be judged criminal.

FUNDAMENTAL VALUES

Central to the foregoing arguments and to related ones throughout the Working Paper is the focus of the Criminal Code on protecting and underlining society's *fundamental values.* "The major focus of criminal law (and only criminal law) is on the highlighting and protecting of the fundamental values of our society by providing in the most serious, emphatic and onerous way available for the repudiation and deterrence of those who have threatened or might threaten them" (p. 43). Gross environmental pollution is to be repudiated and deterred because, by seriously damaging or endangering the environment, it seriously contravenes a fundamental value "which we will refer to as the *right to a safe environment*" (p. 8). This value, the Commissioners say, may not as yet be fully emerged or universally acknowledged, but its existence and shape are already largely discernible. "In protecting it, the Criminal Code would be essentially reflecting public perceptions and expanding values traditionally underlined in the Code—the sanctity of life, the integrity of persons, and the centrality of human life and health" (p. 13).

Such statements, along with the conclusion that "the scope of a Criminal Code offence against the environment *should not extend to protecting the environment for its own sake* apart from human values, rights and interests" (p. 15), clearly show the basic proposition underlying the arguments: namely, that human life and health should be safeguarded. Fair enough. But then is not the title, "Crimes Against the Environment," inaccurate? It should be "Crimes Against People Via Their Environments."

This point is central to the analysis that follows. Crimes against people ought not to be confused with crimes against the life-giving ecosystems that envelop all organisms including people, for two different categories of ethical concern exist. Insofar as each is valued intrinsically, fundamental values attach to each. A close analogy is a "crime" against a late-term fetus as distinct from a crime against its environment—the pregnant woman who carries it—for here are two distinct, though closely interrelated, objects of value and ethical attention.

Much like the infant *in utero,* humanity is brought into existence and then is sustained by a surrounding ecological system that has intrinsic value in the sense that without it there would be no life of any kind, no people, no talk about their rights. In the words of the Canadian Environmental Law Research Foundation, "without an environment capable of supporting the human race, all other rights are useless" (p. 13, footnote 19).

Nevertheless, the Working Paper's rejection (except in the title) of the idea that there can be crimes against anything but people seems at face value to have advantages. Society is prepared to buy the proposition that environmental protection is justified as an extension of human rights to life and health. The idea is easy to understand, set squarely as it is on traditional homocentric philosophy. Pollution is identified as the problem, tied to "the ultimate concern and basis [in] human health" (p. 9). The boggy ground of "victimless crimes" is avoided and, in effect, class action against polluters is proposed for legitimization and formalization within the Code.

The disadvantages are deeper and not so readily discerned. In essence they have to do with perpetuation of the focus on people-centred values in a planetary ecosystem that is rapidly deteriorating just because of them. Species self-centredness is no more likely to save society's sustaining milieu than a narrow obsession with fetal salvation will safeguard the women who conceive, carry and sustain the unborn.

Nor is it certain that the tying of environmental rights to human rights will command universal support. Arguments that people possess a basic right to a quality environment are bound to evoke criticism from those who perceive attempts to widen the circle of fundamental rights as detracting from and threatening the traditional core of inherent human rights. Here a sorting out of different kinds of "rights" is helpful.

FUNDAMENTAL RIGHTS

As is natural for a self-aware species, people value themselves above all else. The most highly valued things have rights, and the attachment to themselves of rights is the obverse side of fundamental valuation of themselves.

Two kinds of rights are recognized: those of life and liberty—intrinsic, inherent, arising just from the fact of belonging to the human race—and those that are conferred and/or acquired as a consequence of merit, work, gift, purchase, inheritance. The first are Human Rights or Natural Rights, the second Moral Rights.[4] Both kinds are recognized by society and are backed to various degrees by the law. But only Moral Rights *depend* on the law for their legitimization. Human Rights, being assumed as intrinsic, require no such legitimization; they are axiomatic, considered to stem from nature, from natural law. "We hold these truths to be self-evident. . . ."

Those who have deep interests in Human Rights tend to look askance at the lengthening list of Moral Rights, claimed by their proponents to be basic. The entire enterprise of protecting Human Rights, say their advocates, is being diluted or undermined by the steady accretion of more and more Moral Rights—for education, for health services, for fair pay, for holidays,—each pronounced "fundamental." The few essentials, they say, are being sabotaged by those preaching various social *ideals* as basic human *rights*.

Will the Working Paper's proposition—that a safe, quality environment is a fundamental societal value and right—stand up to these critics? Only with difficulty. If the reason for protecting environment is only that its deterioration impinges on human health, why so do unemployment and poverty. If those who endanger people's health by grossly polluting the environment are to be prosecuted, should not those who threaten human health by closing down the factory and creating gross unemployment also feel the law's bite?

Crimes against the environment based on a newly discovered people-right to a safe environment will surely be criticized as yet another watering down of efforts to achieve a global consensus on essential Human Rights. And this criticism has a certain logic because, from the perspective of most people (uncontradicted by the Working Paper), a safe environment is only an ideal, a Moral Right to be worked for and possibly achieved sometime in the future, a secondary consideration not on the same footing as, say, freedom of the individual.

Entangling environment with human rights is a prescription for inaction and frustration. Cascading global problems in air, water, soil and food suggest that we cannot afford such an unsure, time-wasting strategy. A faster track to environmental protection is needed. Conceiving environment as having its own intrinsic values immediately arms it with fundamental rights.

UNDERSTANDING ENVIRONMENT AND ITS INHERENT WORTH

Questions of who and what possess fundamental values, and spinoff rights, are increasingly in the public consciousness. Answers depend in part on ecological understanding of relationships—between people, between people and animals, between people and the global environment (the Ecosphere) from which all evolved and by which all are sustained. A critical examination of the ecological understanding reflected in the Working Paper reveals the source of its shortcomings in its attempt to protect "the environment."

Notable in the first place is the absence of any explicit definition of *environment*. Legal experts are not alone in assuming that everyone knows her/his milieu and understands exactly what it is that suffers when "environment is mistreated." Because the report's primary concern is with pollution as a crime, the reader will infer that by environment is meant various disconnected things—the air, rivers, lakes, oceans, soil—in short, the various media that receive health-threatening toxic substances. Also, in several statements, environment is vaguely linked with natural resources (p. 15).

The authors appear to be getting warmer when they use the term "ecosystem," notably in discussion of incremental damage, destruction or injury to the environment (p. 20–25). Only once, however, do they hint at the conceptual equivalence of environment and ecosystem by mentioning the two together: "everything in the environment or individual ecosystems is related" (p. 22).

Ecosystem is defined conventionally as "a unit of nature in which non-living substances and living organisms interact with an exchange of materials taking place between the nonliving and living parts," and examples are correctly given as "units of land along with the surrounding air and water . . . or the Earth itself or the biosphere (the outer sphere of the Earth inhabited by living organisms and including lakes, oceans, soil and living organisms, including man.)" But after this excellent start, the implications are not developed. Instead, a straw-man called "the Unqualified Ecosystem Approach" is erected (p. 22) to be knocked down in support of the truism that "a limited degree of damage to elements of the environment is inevitable, beneficial and even natural" (p. 28).

In ecological science an "ecosystem approach" does of course exist as the logical extension to the world of the ecosystem concept. In the Working Paper, "the Unqualified Ecosystem Approach" is interpreted as an extreme statement of the preservation ethic; namely, that *all* tampering with nature is to be strictly avoided. The ecologists to whom reference is made—Odum, Commoner, Lebreton, Schnaiberg (p. 21, footnote 23)—do not subscribe to this narrow interpretation whose emphasis distracts from the essential insight of the ecosystem concept: that all life participates in a complex and interactive system where boundaries between living and non-living are vague or non-existent. The largest ecological system, the Earth Ecosphere, is the reality of which people are one part; they are embedded in it and totally dependent on it. Here is the scientific source of the environment's intrinsic value.

Part of the problem in the Working Paper is confusion between *ecology as a science,* that uses an ecosystem approach to explain and solve environmental problems, and *ecological philosophy* (ecophilosophy) that explores the implications of human ecology. Of the latter, two streams are distinguishable. The one, sometimes called *Environmentalism,* is homocentric and assumes that nature is valuable only insofar as it is useful to the human species. Its faith is in science and technology for control and repair of environmental deterioration. The other, unhappily named *Deep Ecology* (better *Ecosophy),* is radical in extending high value to the non-human world. It proposes control of environmental deterioration by establishing a new Ecosphere-people symbiosis, by practicing an ethic that transcends humanity.

Apparently the second philosophical position has been equated with the "Unqualified Ecosystem Approach," whose proponents are said, in the Working Paper, to be "unjustifiably pessimistic" (as they are about the likelihood of success of Environmentalism) and "too rigorous" (in pressing the realism of the ecocentric view).

Yet a synthesis is possible. The people-centered or homocentric Environmentalism position adopted in the Working Paper is not so much at odds with Deep Ecology's broader ecocentric view, as it is subsumed by the latter. The reasons are illuminated when the three-dimensional ecosystems that comprise the source and matrix of life on Earth are substituted for the inexplicit term "environment." Then we perceive that people of intrinsic value dwell within an Ecosphere of intrinsic value.

ENVIRONMENT AS ECOSYSTEMS

The planetary environment is the Ecosphere, literally the Home-sphere, the life-filled shell encircling the globe. Comparisons with the lifeless planets show the improbability of the Earth's interrelated atmospheric, oceanic, soil and sediment components, for they are marvellously tuned to the needs of the organisms they support. In the four and one-half billion years since the planet was formed, the organisms have entered into the composition of the environment as it has entered into them. The characteristic oxygen of the atmosphere, the phosphorus of sea water, the calcium carbonate in soils—all are biogenic, produced by life and, in turn, supporting life. No sharp line can be drawn between the living and non-living components of an ecosystem because they are equally important parts of the whole. The world ecosystem evolved as an integrated dynamic unit, of which that late comer—*Homo sapiens*—is one conscious fragment.

From the evolutionary point of view, the Ecosphere is a creative entity. Improbable in composition,[5] it has produced millions of life forms in addition to the human. The miraculous two-layered global environment—a stratum of mixed gases overlying a stratum of water, sediments and soils warmed by sunlight—somehow, over eons, gave rise to people, not the reverse.

A priority of importance is therefore established, further support for which is given by Ludwig von Bertalanffy's level-of-organization insight.[6] Just as each person consists of parts—organs, tissues, cells—that are lower levels of organization, so hierarchically, in the other direction, each ecosystem represents a higher and more inclusive level of organization, an evolved three-dimensional air-water-soil matrix that encapsulates organisms as parts. Humans, important in their own right, are nevertheless a lower level of organization than the Ecosphere and its sectoral ecosystems within which, with other plants and animals, they are functional units.

From this perspective the claims of people to possession of the Earth would be amusing were they not so threatening. Evolutionary ecology lends little support to the quaint notion of a dimension of common ownership expressed in the Institutes of Justinian: "By natural law the following things *belong* to all men, namely: air, running water, the sea, and for this reason the shores of the sea" (quoted p.12, emphasis added). By any reasonable rules of priority and descent, the labile parts of the Ecosphere (atmosphere and hydrosphere), plus the more stable continental parts to which the claims of private property are made, do not belong to humans; humans belong to them. Land is a community to which we belong, said Aldo Leopold; not a commodity that belongs to us. In a profound sense, the world environment is not our heritage; we are its heritage.

HOMOCENTRIC AND ECOCENTRIC LAW

The Commission has concluded that a Criminal Code offence against the environment should not extend to protecting the natural environment for its own sake. How reasonable is the position that links environmental protection strictly to human values, rights and interests?

Essentially the Commission's arguments hinge on the proposition that only people are important, fundamentally valuable and naturally gifted with rights. Nevertheless the Commission has not closed the door on the possibility of recognizing rights of non-human objects "at some future date" (p. 10); it just does not think that now is the appropriate time for radical action. At the same time, looking to the future, the informing role of the law is also recognized. The Code can play an educative and advocacy role "by clearly articulating environmental concerns and dangers not always perceived as such" (p. 15).

The Commission's case against extending protection to environment per se, essentially denying its intrinsic worth, is summed up in five propositions. My commentary follows the statement of each.

(1) *Such an extension of law would be revolutionary, not evolutionary.*

Here is a call for gradualism, for conservatism, for slow evolutionary changes in the law. Such incremental changes in social institutions, as in organisms, are the normal response to slow environmental changes. Unfortunately the problem today is one of rapid deterioration of world ecosystems due to exponential population growth, energy use and industrial development. The expectation that incremental evolutionary changes in social institutions will be able to cope with the explosive causes of environmental degradation is unrealistic. In biological terms, social responses must be of the macro-evolutionary type, saltatory or punctuative, not incremental. A paradigm change, a radical shift in the way that people see themselves in relation to the planet, is necessary. The law should not be timid in expediting changes that require a radical break with the past.

(2) *It is inconceivable that natural resources (i.e., environment) could ever be totally insulated from economic and political considerations.*

The implication that care for the environment means insulating it from people and not using it is a red herring. All organisms necessarily live by and from the Ecosphere. Valuing the planetary environment for itself neither means its non-use nor proscribes every form of "pollution," if that means the release of unhealthy wastes. It *does* mean directing resource use in conserving rather than exploitive ways, maintaining genetic diversity and essential ecosystem processes.[7]

Humans cannot escape being heterotrophs—animals whose economic needs require access to the world's energy and material resources. But recognition of this fundamental dependency should raise the value of the life-sustaining milieu to that of an end even as it constitutes the means by which humans are sustained. Some of the world's ecosystems should be strictly preserved, and their components should be given names other than "resources" to indicate that they are sacrosanct. Other ecosystems must be used in the way that all animals use their environments, but with a caring attitude. The attitude will make the difference.

(3) *It would amount to granting rights to non-human entities, and it has always been supposed that only humans can have rights.*

This expresses the ancient, ecologically naive tradition of people valuing only people, supported by a homocentric ethic passed down from a long humanistic history in which environment-as-nature has been viewed with distrust as an alienated world, as less than human, as inferior and unworthy of fundamental valuation.

Earlier in this essay, in the comments on Human Rights and Moral Rights, the linkage of rights and values was stressed. Fundamental rights—Human Rights—are the expression of perceived high value; they are not conferred or otherwise acquired but simply recognized, out of a sense of rightness. Therefore, nothing but ignorance of the importance of the Ecosphere stands in the way of acknowledging natural environmental rights. The problem is not that of granting rights to non-human ecosystems and the Ecosphere (as the State grants Moral Rights to its citizens) but of the *recognizing* the rights of these supra-organismic entities, just as the inherent worth of people is justly recognized in their rights to life and freedom.

(4) *Efforts to argue the case for recognition of environmental protection for its own sake have not gained general support in philosophical or legal thinking, and some very real problems stand in the way unless consideration is given to human benefits, wishes, uses and health risks.*

This is perhaps the weakest part of the Working Paper's rationale. The "very real problems" are not identified; they are simply bypassed with a footnote reference to three papers: C.D. Stone's, *Should Trees be Standing?*;[8] L.H. Tribe's, *Ways Not to Think About Plastic Trees*;[9] and D.P. Edmond's *Cooperation in Nature: A New Foundation For Environmental Law*.[10] These three authors do mention conceptual problems, but they also point toward the solution in recognizing environment's intrinsic values and rights. They might agree with the Commission that efforts "to assign rights to nonhuman entities . . . have so far not met with anything approaching general support" (p. 10), but this is exactly the attitude they deplore as they search for the means to change it.

Stone recommends the "unthinkable" conferring of legal rights on natural objects, following the precedent that rights are afforded to such legal creations as ships, municipalities, corporations. Additionally, he says Nature has rights on its own account. Tribe also moves toward recognizing environmental rights per se, remarking that nature embodies values apart from its usefulness in serving man's desires. He suggests avoidance of the premise of human domination, encouraging a stance of criticism toward current people-environment attitudes, and commitment to the conscious improvement of the world environment.

Emond looks askance at conferred environmental rights but only because he fears that they will not stand without the foundation of a new social paradigm; namely, that of co-operation and mutual aid replacing the current rule of competition and individual aggrandizement. Although he notes that "Rights may derive from the thing itself in the sense that certain rights spring from our humanness," he does not pursue the subject of intrinsic rights and its implications for the environment. He takes the legal positivist's position that all "rights" derive from human society. All are conferred, in the same sense as Moral Rights.

Emond is pessimistic because of the untrustworthiness of conferred rights that posit a dominant-subdominant hierarchical order of importance, for what strength can such rights command when the donor (society) knows itself to be more important than the recipient (environment) by the very fact that the one gives as benefactor while the other receives as debtor? Without a new way of seeing the world that makes people and nature symbiotic collaborators, equal in co-operation, the extension of rights to environment is unlikely to be helpful.

Implicit in Emond's argument is the goal of recognizing environment's intrinsic rights, for only when the two "objects" of importance—humans and the world of nature—are valued in and for themselves can they approach equality. Only then will the "unreasonableness" of environmental protection and preservation "for its own sake" appear reasonable.

All three writers in their different ways place high value on the world environment, and all support radical legal approaches for enhancing its rights, conferred or intrinsic. The nub of their common problem is Western culture that assumes the most important human right, after the right to life, is individual freedom. This is the heart of classic liberalism, derived from the ideas of the Enlightenment and expressed in the writings that flowed from the American and French Revolutions. The dilemma is that without the context of a higher good, freedom for people leads to the enslavement of nature.

(5) *Recognition of intrinsic environmental rights is unnecessary. The homocentric ethic can provide protection for the natural environment by giving*

greater emphasis to quality of human life as a goal and by more attention to the responsibilities of stewardship and trusteeship.

The seriousness of the planet's deterioration over the last half century suggests that the homocentric ethic is itself the problem. More efficient housekeeping, better stewardship and trusteeship, is unlikely to help because the ethical source of mistreatment of the planet—human species self-centeredness—is untouched.

The French *Declaration des Droits de l'Homme et du Citoyen* defined liberty as "being unrestrained in doing anything that does not interfere with another's rights." In line with this popular sentiment (that today would see property rights enshrined in the Constitution), George Grant defined liberalism as the set of beliefs proceeding from the central assumption that man's essence is his freedom, and therefore what chiefly concerns man in this life is *to shape the world as he wants it.*[11] Here is the prescription for the massive environmental destruction that is evident wherever Western culture's influence is felt, destruction whose motivation only the recognition of nature's intrinsic values and rights can overcome.

The legal dilemma is posed by the conflict of values in the first and fourth "signposts" of Code crimes set out earlier in this article. Offenses should be considered real crimes only if:

(1) they contravene a fundamental value;

(4) the needed enforcement measures would not themselves contravene a fundamental value.

Were society to recognize the intrinsic and fundamental values of the Ecosphere, then numerous environmentally destructive actions beyond those that threaten the vital needs of *people* would qualify as real crimes. The enforcement measures against the human perpetrators of the crimes might well contravene the fundamental human value of liberty, of freedom "to shape the world as we want it." When values clash, which possesses the highest value and takes priority over the other—that of the Ecosphere or that of the individual and society?

Ideally, Emond's philosophy of co-operation and mutual aid

could soften the people-Nature conflict. In the crunch, however, the Ecosphere (Nature) ought to be valued above people on the basis of precedence in time, evolutionary creativity and diversity and the complexity of a higher level of organization.

Conceivably, for example, present human population, expanded in size by technology, has become an active evil, exceeding the sustainable limit, overwhelming the planetary environment. The ultimate crimes against the environment, crimes that also threaten the human enterprise, are fecundity and exploitive economic growth, both encouraged by the homocentric philosophy.

The argument in this paper for moving from the homocentric to the ecocentric ethic is an argument for broadening values. It does not deny high human valuation; it adds to traditional humanism a deepened appreciation of the surrounding world with which people *must,* sooner or later, work out a cooperative symbiosis. The alternative, the species-centred convention that rules today, is stifling. In the words of Tribe (op. cit.): "the homocentric logic of self-interest leads finally not to human satisfaction but to loss of humanity."

CONCLUSIONS

Laws both guide and contribute to public consensus. Today's environmental problems, global in scale, call for much more than a reactive role with incremental change. Arguments for micro-reform of the law are mismatched to the challenges of macro-environmental deterioration. Recognition by the law of intrinsic environmental values and rights could be a powerful influence, facilitating a saner and safer Planet-people relationship.

Attempts to base environmental protection only on people's rights (the right to a safe, attractive environment) risk the danger that the latter may be disputed as just another of the lengthening list of ideals that distract from basic Human Rights. Although the Commission may pronounce the right to a reasonable quality of environment as "fundamental," its status will not thereby be raised above any other conferred or Moral Right as promulgated, for example, in the United Nations' *Universal Declaration* of 1948. Such rights do not command high priority. As the Experts Group on Environmental Law to the World Commission on Environment and Development reported, "the fundamental human right laid down in Article 1 (to an environment adequate for human health and well-being) remains an ideal which must still be realized."[12, 13]

Again, if the right to a safe environment is espoused as fundamental *because* of the dangers that unsafe environments pose to life and health, should not other rights (to employment, to income, to housing) be equally recognized as "fundamental"? The effect is to slight rather than emphasize the overriding importance and value of the Ecosphere.

At the moment, environmental law suffers from the lack of a strong conceptual and normative basis. The deficiencies are perilous, for in effect they forestall the kind of rational action needed to counter the universal problems of a life-threatening deteriorating global environment. The only way that the human race can do itself in, finally and completely, is by contributing to the destruction of the planetary Ecosphere, either incrementally by pollution, toxification and erosion, or catastrophically in one big bang using nuclear technology. Hence the call for clear ideas about the meaning of "environment," about its fundamental significance, values and rights.

The Law Reform Commission has underlined the need for expansion of the rule of law into those social situations where the pressures of population and industry depreciate and destroy the quality of the Earth's interrelated air, water, soil and biota. Labouring under serious conceptual difficulties, it diagnoses the fundamental problem as pollution that affects people.

The focus remains fixedly on bodily harm while toxic wastes accumulate, soils are degraded, forests poisoned, plants and animals exterminated, natural ecosystems slashed, burned and bulldozed into oblivion and ground water resources destroyed.

It is time to shift the focus, recognizing the surpassing values of the Ecosphere and the seriousness of crimes against *it*.

ROLE OF THE UNIVERSITY

Years ago the university shaped itself to an industrial ideal—the knowledge factory. Now it is overloaded and top-heavy with expertness and information. It has become a know-how institution when it ought to be a know-why institution. Its goal today should be deliverance from the crushing weight of unevaluated facts, from bare-bones cognition or ignorant knowledge: knowing in fragments, knowing without direction, knowing without commitment.

Ignorant knowledge is exemplified by the remark attributed to Enrico Fermi as he contemplated the unleashing of atomic power: "Don't bother me with your conscientious scruples; after all, the thing's superb physics!" Engrossed in the particularities of physical science to the exclusion of all else, Fermi and his co-workers were not by that fact socially detached nor acting apolitically. They were initiating, in ignorance, a sinister drift into nuclear nationalism.

Various scholars have drawn the distinction between mere knowing how to do things, how to make things, how to accumulate facts and figures, and evaluative thinking that prepares one to judge what ought ethically to be done.[1] Knowledge without evaluation seems a safe pursuit, unthreatening because it simply adds to what has proved useful in the past. In the long run, however, and particularly in times of change, it is dangerously unconstructive because it has nothing to say about choice of directions. Evaluative thinking, seeking know-why rather than know-how, recognizes the need to act ethically in society and the world. It is constructive, although often unsettling and dangerous to the established order.

The distinctive role of the university, in a volatile society set in a deteriorating world-environment, ought to be the quest for know-why ahead of know-how. My thesis is that the university should be giving society what it needs most: constructive criticism, cultural understanding and ethical alternatives on which to act. In the words of James Downey:

> The essential purpose of the University is not to carry out research, nor even, in the conventional understanding of the term, to teach, but to furnish a critical commentary upon the assumptions, beliefs, values, knowledge, and technologies that inform and support the social order.[2]

In the same issue of the journal that carried Downey's thoughtful piece, Larkin Kerwin of the National Research Council asserted that *research* is the heart of the university. Who needs critical commentary?

UNIVERSITY AND KNOW-HOW

Judged by the criteria of critical commentary on the social order and of concern for the directions and goals of society, the North American university earns low grades. Its performance is inadequate and uncreditable. In the various colleges, particularly but not uniquely the applied or "professional" colleges, far too much emphasis is placed on unreflective knowledge, on the collection and filing of facts and all manner of know-how. The synthesizing search for intelligent and ethical know-why is neglected. The university is content to travel first class, on a fact-overloaded boat without a pilot, full steam ahead, in a fog, not knowing where it wants to go and not asking.

Obvious risks attend criticism of any organization as large and diverse as "the university." The institution has in fact become a multiversity, comprising a heterogeneous mix of students and faculty who profess all manner of beliefs and opinions. The attribution to it of purposes and goals may seem questionable. Nevertheless, institutions function from year to year on the basis of conventions that reflect an underlying consensus on roles and goals, and these can be scrutinized, evaluated, criticized.

If we stand back and look clear-eyed at the departmentalized faculty and the discipline-fragmented curricula of Canadian universities, then the structural and functional obstacles in the way of acquiring integrated what-to-do knowledge are seen to be pervasive and dominating. And if, as sharp as Aristotle, we believe that human nature does nothing in vain, then our suspicion is justified that this institution of higher learning is intentionally geared to a fragmented world-view in order to keep it safe, serving business-as-usual, rather than society-as-it-might be.

The boast that the university is operating at the cutting edge of research and scholarship conceals the ethical vacuum at its core. It prides itself on being up-front, leading the way, delivering to society what it needs most: the know-how of specialists. Name any subject; the university has its experts. Taken all together, they must have all the answers. Believers in the virtues of compiling information and know-how must subscribe, however, to an irrational faith in a benign "unseen hand "that will guide the bits and pieces of atomized knowledge into an harmonious whole for the good of all.

Equally irrational is the related "temple of science" faith: that each researcher—no matter how lowly and obscure—by adding her brick of new information to the pile will somehow participate in the erection of a noble edifice. Why the fortuitous result should not at best be a scenic ruin, and at worst a jail or death camp, is for most researchers unconsidered and unimaginable.

DECLINING FAITH IN PROGRESS

The god in the machine that has long been trusted to save us from the results of ignorant knowledge is Progress. For the true believer, Progress moves society under the hand of historical necessity toward unknown but desirable goals. In the brighter days of the 1920s, the philosopher Alfred North Whitehead, with an eye on both old-world and new-world universities, spoke of them as the suppliers of intellectual imagination in the service of progress. Their existence, he said, is the reason for the sustained rapid progress and progressiveness of European life in so many fields of activity.[3] He did not define progress and progressiveness; he did not need to. Everyone knew them to be the soul and inspiration of the expanding industrial society.

In the early years of the century, the orientation of all Canadian universities was straight and sure because of the overwhelming consensus that Western industrializing civilization was on the right and only track. It had been determined that Truth and Progress were on the same standard and together they flashed a beacon up ahead. The university role was to speed society toward the light. Campuses, proud of their secularity, sprouted buildings in the style of Gothic cathedrals, reflecting a devout conviction that glorification of the Arts and Sciences within the technologic milieu could not fail to bring blessings to humanity.

Persistent doubts about Progress appeared after World War II and more so after the ecological awakening of the 1960s. Today enthusiasm for progress is muted, as are its retinue of stirring concepts—a better world, a more peaceful world—on which earlier generations of students were raised. Yet surprisingly, society has not been deflected from its mid-century direction. The mode is still acceleration down the same old track, even though the signs as they flash by are less and less encouraging and the leading beacon has gone out.

UNIVERSITY GOALS

One symptom of disillusionment is a retreat from using language that implies society knows where it is going. Thus universities assess their contemporary role vaguely but with studied elegance. Eminent academics from the international community provide examples in a recent series of articles on understanding the modern university.[4] The heart of the university, says Eric Ashby, is the partnership of teachers and scholars engaged in the detached and disinterested study of intellectual systems. "Detached and disinterested" places the university squarely in the know-how field, blunting its potential for leadership. Ashby quotes Alvin Weinberg approvingly to the effect that, at the university, the specialist and analyst is king. Which seems too bad for society, for the loyal subjects thirsting for the wisdom of engaged synthesizers and thoughtful non-specialists.

Other contributors float a variety of abstract ideals. John Gardner defines the university as that extraordinary institution from which come the deeper insights and understandings that keep our civilization alive and humane. It is, he says, committed to the idea of individual worth and dignity, and it can call us to the best in our tradition. Does "the best in our tradition " measure up to present needs or should we be attempting a new start?

T.R. McConnell draws on many sources to argue the university role as cultural enlargement through the reconciliation of reason and feeling. Mina Rees calls the university the locus of the ivory tower through which intellectual discipline is transmitted in the quest for truth. For Brand Blanshard the university is the citadel of rationality in thought, feeling and practice that supports and promotes the reign of reason. Most at home in modern academe, David Gardner recites the wonders of electronic gadgetry that will, he believes, add to the effectiveness and economy of higher education. The medium is the message and content be damned?

Similar sentiments are expressed by spokesmen for Canadian universities. Especially at convocations, the clichés are paraded. Universities are centres of excellence in the pursuit of truth, institutions to sharpen minds, places for education in democracy whose essence is to provide an intellectual experience, and so on. Typical is the mission of the University of British Columbia. It is:

> like that of all first-class Universities, to serve society by providing the best environment it can in which to nurture and stimulate the native intellectual capacity, curiosity and creativity of its students.[5]

Bland phrases about nurturing minds and developing intellectual resources eclipse the aims of education. They inspire no challenging visions of the future nor do they dispose anyone to joyous action, for they fail to make any connection between learning and living. The words convey one subliminal message: seek not the end of knowledge; have faith that it is enough to know.

We should remember the University of Chicago whose motto is: "Let knowledge grow from more to more/That human life may be enriched," and Milton Mayer's comment that by 6 August 1945, the University's knowledge had grown to the point where it was able to enrich practically all the life of Hiroshima.

AVOIDING THE KNOW-WHY QUESTION

Various circumstances encourage academics to avoid confronting the knowledge-for-what question. Those who habitually work with abstract concepts run the risk of losing contact with the world, of becoming misty or psychotic, and this is an occupational hazard for lifelong commuters to the ivory tower. But there is also a widespread enervating sense of hopelessness abroad, because the future has been glimpsed and it is repellent.

The Western dream, that once gave purpose by promising progress, bids fair to turn into the nightmare reality of a totally mechanized and robotized world from which all things organic—except those genetically engineered—have been eliminated. Next-Year Country is Orwell's 1984, with the additional threatening dimension of global annihilation.

Unwilling and perhaps no longer able to ask the question "Knowledge for What?", the university has accepted as its primary focus the stripped-down cognitive function. Knowledge for the sake of knowledge, "detached and disinterested" in the words of Ashby, is elevated to a place of honour and prestige beside Art for Art's sake. The more oppressive the world becomes outside the hallowed halls, the stronger the incentive to make a fetish of cognition and stick to disseminating knowledge. The quiz kid mentality is encouraged, and to be educated is equated with having a fact-filled head. The campus capitulation is complete if we take at face value the Preamble of the Collective Agreement of one Western university: "The parties recognize that the goal of the university is attainment of the highest possible standards of academic excellence *in the pursuit and dissemination of knowledge . . . .*"[6]

It bears repeating that fragmented knowledge, assiduously pursued and disseminated, is not neutral. It fits smoothly into and greases the gears of a mechanistic world view that has brought undoubted material benefits to society—the early progress of which Whitehead spoke. But now that same mechanistic world view, along with the science-technology it has spawned, has taken on a monstrous life of its own. Unrestrained and out of control it devours the Ecosphere and threatens to dispose of humanity by slow poisoning or in one last big bang.

For observe that know-how, the unreflective part of rationality, is a means that in itself provides no end save that of power. Like all other tools, knowledge (as information and expertise) is power, conferring economic advantage and technologic control on those in a position to use it. When distributed democratically in fragmented form—and we live in a veritable blizzard of facts and information—it clouds the social vision and obscures the questioning of directions, thus strengthening entrenched privilege. Merely adding to the stock of such knowledge without any assessment of its world/life-enhancing or world/life-destroying potential is certifiably blameworthy if not criminal. Unguided by ethical understanding, crude know-how creates national and international instability and poses escalating dangers to humanity and to the Ecosphere.

No country should be satisfied with a merely literate, information-knowledgeable and computer-wise citizenry. To evolve in a civilized way, Canada needs social critics, grounded in the culture, understanding from a world/life-enhancing perspective the potential values of the Arts and Sciences. The university should be a congenial and supportive home for such people, and their nurture should be its highest priority.

Furthermore, university structures should reflect the understanding that knowledge can be turned toward wisdom when it is comprehended as whole, cross-disciplinary, value-oriented, ethical. Courses in applied ethics should not be just the narrow peripheral interest of medicine, commerce and law; they should be the core of all curricula, enlightening academic endeavours. A foremost concern of faculty should be the orientation of knowledge to right action, guided by the old-fashioned but indispensable ideals of equality and justice as well as by the new-fashioned insights of ecological harmony, conservation and attention to securing a sustained environment.

But what in fact concerns the university most? Lack of buildings, lack of classrooms, lack of laboratories, lack of research equipment, lack of experimental animals; in short, lack of physical plant. To be fair, concern is also expressed about shortage of faculty, though whether quantitative relief of that problem would effect qualitative improvement is a moot point.

The university campaigns vigorously for resources "to keep faculty at the frontiers of science"; it does not ask what these frontiers offer in the long run, nor indeed if they may prove inimical to humane living. The university "strives to keep abreast of technological change"; it expresses little concern that the technologic juggernaut on its current course may well destroy the world.

The question "Knowledge for What?" is not asked because, to repeat Weinberg's dictum, at the university the specialist and analyst is king. On this, the humanities and the sciences, basic and applied, are in complete agreement, and, according to this consensus, the university's academic rewards are annually dispensed. No particular merit is attached to inter-disciplinarity, and, in the review process for promotion and tenure, the faculty member who explores the significance of his/her field by transgressing disciplinary boundaries is judged a "Special Case," a label that marks such true academics as egregious oddballs, to be tolerated, perhaps, but certainly not emulated.

SPECIALIZATION OR CULTURAL UNDERSTANDING?

Specialists and analysts in the university reproduce specialists and analysts as graduate students, sending them forth as trained tinkerers, ill-prepared to be thinkers-in-action. They score high on means tests but low on ends tests, for they are not equipped to be social synthesizers. Sorting out goals is difficult for those who have never been confronted with the key kinds of questions. Little wonder that feelings of inadequacy are solved by reciting The Creed taught at the university: "Our role is to provide specialized, impartial, objective knowledge; what society does with it is society's problem." A great deal that is pernicious in society battens on such naiveness.

I do not criticize the university for abstaining from direct social action but for neglecting intellectual activity that is oriented to social action. Nor do I dispute the university's role as that of educator in a monastic setting, devoted to the pursuit of great ideas in a quiet haven withdrawn from the hurly-burly of real life. The Gothic and Classical buildings, the enclosed gardens, the avenues of trees, the quiet walks are appropriate to the contemplative life that precedes and is preparation for the world of action.

Yet knowledge is inseparable from the activity of living because it is an artifact of culture. What we see and think and know depends on when and where we are born, on where we happen to be in time and space: hence the responsibility to understand the contemporary scene and its motivating values and ideas to the best of our abilities so as to participate in the flow of cultural existence. Why then should the university community set such high value on the disengaged and dispassionate search for truth, as if social directions were the business of other institutions? Is higher education only academic-cerebral?

No, for neither science nor scholarship stand apart from societal goals and their value-laden content. Science is the culturally conditioned pursuit par excellence, shot through and through with Western suppositions concerning the primary reality of a material world and what is good to know about it. The questions that researchers and scholars ask and the solutions they take to be true must conform to certain paradigms and theories neatly nested together in a world view that is an outgrowth of Western culture. All activities in the sciences, as in the arts, are the fruits of a particular historical-social evolution that runs strongly, in our tradition, to hierarchy, domination, paternalism, control, utility. Inherited thought-systems direct attention to particular kinds of "truth" consistent with the tradition.

What cries for examination and critical evaluation today is not the truth-value of the factual information and knowledge that deluge us, for they find their justifications in the paradigms of society. It is these pat paradigms themselves that need to be questioned: the various "unseen hands" and other faith objects to which we have surrendered critical thought.

What is the truth-value, as judged by their life-affirming and harmonious effects, of the fundamental belief-systems under whose guidance and in whose context society pursues knowledge, makes political decisions, erects new laws, trims the economy, explores the Arts and bankrolls Science? Can it be that the paradigms on which, knowingly or unknowingly, Western civilization has staked its life, as well as the life of the rest of the world, are dead wrong? Can it be that they are, in the long run, life-denying, fostering sickness instead of health? The media, especially their business sections, are littered with failed diagnoses of basic human problems just as the environment is littered with evidence of death-wish technology, but still homage is paid to bankrupt traditional beliefs.

The New University might well adopt as its ensign, inviting continuous questioning of its directions and goals, the lines by T.S. Eliot:

> Where is the wisdom we have lost in knowledge?
> Where is the knowledge we have lost in information?[7]

ETHICAL ECOSPHERE

Ethical problems are problems of conflicting values. We value things, ideas, beliefs, faiths, with varying intensities. When values held with similar ardour clash, then a serious ethical problem emerges. In the arena of environmental ethics, a high valuation of people competes with a rising valuation of the Home Place.

Some people associate ethics and morality with the narrow mind and manners of Mrs. Grundy, thinking of codes, rules and regulations that cramp and confine. Others equate ethics with social controls over business finagling, medical chicanery and sharp legal practices. But to disparage morality and ethics as no more than systems of prudish or protective “dos” and “don’ts” is to ignore their roots in the fundamental values that evolve at the centre of human culture. They are connected to the springs of human motivation and will change as root values change.

The urge to understand ourselves and our world merges ethics, art and religion, for all are in pursuit of the same goal: new valuations befitting new realities as they are revealed. We know, perhaps too well, the value of ourselves, and an appreciation of our own kind is strong in traditional ethics, art and religion. Can the world at least equally be valued in all three important modes?

An affirmative answer for Nature will not come simply from exhortations to change our ways. We value things and act ethically toward them when we feel their importance deeply and when, just as profoundly, we understand why they are important. An unshakable ethic for the Ecosphere will emerge when we believe in our heart and minds that our worldly environment is a reality more important than me, you and all of us. When such a conviction about Nature becomes second nature, we will know that we are parts of the ecological whole that produced us and sustains us.

Because ethical problems revolve around values and their relative importance, we must examine our values to help reform our ethical sensibilities. Aldo Leopold preached the need for a land ethic,[1] but that new ethic can only come when land is valued. If very little or no importance is attached to the idea of "land," then ideas about land as a physical entity will not make it an object of ethical concern. It will fail the test of being a moral object, and exhortations to respect it will ultimately be ignored. In the crunch, when choices are made, whatever is relatively valueless will be sacrificed to the more valuable. The planet, its land and water and air surface, is being beaten and poisoned to death because, compared to people, it is considered to be relatively unimportant. We have developed little feeling for it as a valuable thing. Nature study has barely made it into our school and college agendas and our natural attraction to the world remains uneducated, diffuse and unrefined.

The key question for environmental ethics, therefore, is how shall we value the Ecosphere? Does it deserve our sympathy and care? What is its importance relative to other things we value? Can we make it a moral object? Certainly its significance is gaining ground as air, water, soil, food, plant and animal life deteriorate, goading us into action, but as an admired thing, a loved thing, it is far from displacing people, animals and plants from centre-stage. George Bush speaks for the majority in conceptualizing a future "where self-determination and individual freedom replace coercion and tyranny, where economic liberty replaces economic controls and stagnation, and where lasting peace is reinforced by common respect for the rights of man."[2] He envisions the triumph of human rights, which superficially is well and good—except that their full flowering may mark the death knell of the planet and its rights. And when no Home Place worthy of the name is left for our species, human rights will be only an empty ideal.

The Ecosphere is degenerating because of our people-first attitude, and a dual problem for environmental ethics is how to elevate the importance of the Ecosphere while putting a damper on the overweening pride and self-aggrandizement that plague our species. To value the Earth more and to value people differently—not less but as an essential collaborative part of it—seems necessary if over-exploitation of the globe is to be stopped. As long as the needs and wants of the people have first priority, we will continue to pummel the second priority—the planet. Relative values set the priorities.

What, then, are values? How are they affixed to objects, and can they be changed? What are the possibilities of people learning to evaluate the Ecosphere more highly, making a supreme moral object of it?

Values seem to have three intertwined sources, two in the observer and the other in the observed.[3] First is the powerful cultural source, comprising all the learned likes and dislikes of particular societies and religions, transmitted by language and other symbols. This first source is primarily important because if an idea, a concept, is not expressed in a culture then for that culture it does not exist. Second is the inherited genetic source, the instinctive likes and dislikes, expressed in the behaviour of infants who, without thought, value warmth, comfort, food, sweet liquids, familiar faces. The third can be called the objective source, comprising all apparent emanations from the thing itself. If the emanations are harmonious and empathic to us, if the resulting experience is of wholeness and health, then our response projects beauty to the thing perceived.

The Ecosphere has not been an object of importance and high value, a moral object of great interest and concern, because our culture has not conceptualized it as such. No commonly used words for it exist, and language reflects our ignorance and neglect. The cultural source of values has therefore been stoppered, also effectively closing the mind to the other two sources. Without the cultural idea, without the elevating concept, a focus for the other two value sources—the instinctive and the objective—is missing. Our instinctive love of nature has been lavished on scenery, on pleasant landscapes, on flowering plants, on birds and other animals, but not on the unsensed Ecosphere. Because we are enveloped in it, the whole has not been comprehended and its objective values of beauty and harmony have largely passed unnoticed.

Change is on the way. Now that the Ecosphere has been seen from the outer-space perspective and described mythically as Gaia, now that it is appreciated by some as a complex and diverse ecological system, its valuation is on the rise. For the first time, we are being invited to love a Nature we can both see and sense, focusing on the reality of the planet's surface film—the living and more-than-living Ecosphere. Ecology is contributing to the sense of importance attached to the world while also indicating the appropriate role of people in the greater whole.

Moral standards and ethical actions are human inventions, tied to beliefs, faiths and understandings. They are our own, formed by us and therefore human in form, homomorphic. This does not mean that they are necessarily homocentric, centred on ourselves. Ethical action need not be confined to our own kind but can be extended to whatever we choose—whenever and wherever we recognize the values and importance of things outside our skins. Specifically, we need not confine moral concerns to those protoplasmic fragments of the world that are conscious or sentient, for such organic parts, though significant, have no monopoly on importance. Biocentrism that limits value-laden concerns to people, to endangered species, to animal rights and to biological phenomena in general is a dangerous detour from the Way—which is valuing the largest unities, the most complete realities that we can comprehend.

Our traditional realities have been Man and God—objects of such value and ethical importance that people have died by the millions in the name of both. Between the two, another reality of importance must be interposed—the Ecosphere. Its surpassing values merit our sympathy, love and allegiance. Already in the radioactive jungles of America and in the chainsawed jungles of Amazonia it boasts a few martyrs. These are the early signs of a New Ethic on the way—for new visions, new values, new religions never arrive without travail. The Way is lighted for us and now another human voyage of exploration can begin, exhilarating and much less lonesome than before.

PRO-WORLD CHOICE

Opponents of birth control, or of abortion as a method of birth control, are prone to use such phrases as "murder of innocent beings" and "war on unborn babies." Is this exaggeration for effect or can such choice of words be justified? It all depends on socially accepted definitions: the definition of a human being and the definition of the time in development when a human being can be said to exist.

To clear away one source of confusion, human life as we know it never "begins." All life today is an extension from the remote past, flowing to the present from origins lost in cosmic time, billions of years ago, passed along from cell to cell by replication in a nurturing world environment. Whatever "life" may be, it is an ever-evolving property of the Ecosphere and its sectoral ecosystems and not just of organisms.

Biological studies indicate that every living cell contains the genetic material to be toti-potent; that is, it contains the potentiality for development into a total organism, a complete adult, given the right surroundings. The environment is essential; it is even more creative than the cells that come from it.

Cells are continually sloughed off from our bodies. Should we consider each living one a person? The potentiality is there just as in a fertilized egg, awaiting only the nourishing environment that some day soon a dexterous tissue-culture biologist will discover and provide. When that breakthrough of dubious value is made, when every living cell of human origin can be encouraged to divide and differentiate in a glass uterus, will all such cells forthwith be pronounced human beings? Surely common sense will say no.

Similarly with sperm and egg cells, industriously mixed in vivo and in vitro everywhere around the world. Are they, and the zygotes they sometimes form, human beings? The majority opinion—for good and practical reasons—is that they are not. Evidence is the fact that although at least as many miscarriages (spontaneous abortions) as live births occur yearly in North America, none are tabulated as human deaths and none are investigated as possible murders. To become a human being requires development in form and function to the stage where a certain degree of independence from the maternal environment is attained.

The difference between potential and actual is basic to the debate on conception control and birth control. An acorn is not an oak tree, although it may become one if placed in the right kind of living system. Tadpoles and roe are not frogs and fish, except in potential. Nor are sperm-egg zygotes and early fetuses people. To call them "innocent humans" and "unborn children" makes as little sense as calling them "unborn octogenarians."

Misuse of language obscures what should be clear: that only a very small fraction of the reproductive potential of any species, ours included, can be realized and accommodated, and that for the fraction of cells fated to become human beings an extended period of organic change and development in a receptive life-giving environment is fundamental.

Living as many of us do in the artificiality of cities, alienated from the biological world, we soon forget that seed mortality—the death of fused sex cells—is natural for all forms of life. Whether and to what extent we should assist the process is an ethical question that should take account of much more than the latent capabilities of the seed.

The potentialities of cells and seeds are elicited by creative environments to which primary ethical concern should be directed. The actualization of good people involves their development in maternal environments that want them and are prepared to accommodate them in healthy ways.

The maternal environment of all humanity is the world, the Earth, now increasingly weakened and ill from too much childbearing. Badly overpopulated and polluted, short on nourishing resources and with its restorative powers crippled, the planet lacks a health care plan and a corps of dedicated healers.

From this perspective, the argument that every zygote and fetus should achieve its potential just because it has that potential, regardless of whether it is wanted and whether its environment can handle it, may in some devious way make humanistic sense, but it is ecological folly.

Considering the tremendous problems that 6.2 billion people are causing, mandatory motherhood seems dubious in the extreme—especially when prescribed by men. The critical questions go well beyond the serious ones of proliferating humanity. Can an urbanized people take thought for the creative Earth environment that sustains all life? Can we escape species selfishness and take action on larger-than-social themes?

How many are ready and willing to make the Pro-World choice?

THE NEW NATURE

Words and phrases are little classifications of experience, summary clips of the meanings of things. They are symbols of the concepts that make sense of our world. Words like *house* or *home* are fairly exact, referring to objects that everyone understands in about the same way. Others like *Nature*—the subject of this essay—have no clear referent; they are diffuse and can mean a whole range of things depending on one's cultural background and education. Such words can generate serious misunderstanding unless they are sharply defined, for when language is loose or defective it can lead us astray.

A critical attitude toward familiar language is important because it channels our thoughts toward the commonplace, encouraging a traditional view of reality. It is conservative, the bearer of past meanings, constraining our reactions in times of changing experience. As feminists know, it frequently perpetuates misconceptions and buttresses outworn moralities. On the positive side, words can be reinterpreted, enlarged or renewed to make them more useful, more appropriate, more realistic.

Today, as environmental deterioration encourages a reevaluation of our relationship to the world, old words and phrases obstruct the way forward. They are awkward in the presence of novel experience and tug our attention back to conventional ideas and modes of thought. Language is a significant part of today's environmental problems and will remain so—until we adopt and feel comfortable with new words like "Ecosphere" and "ecosystem" or with new meanings for the old word *Nature* that are better suited to express and explore the horizons that the new dawn reveals.

The difficulty with conventional language is well known in science. When physicists began to explore atoms and subatomic particles in the early years of this century, they found that light sometimes behaves like a wave and at other times like little packets: quanta or photons. How can a thing be a wave and a particle at the same time? The answer is that the word-concepts of *thing, wave* and *particle* are inappropriate on the atomic plane where no things wave or pulse and where time and space go hand in hand. Waves on a lake and particles of sand on a beach, yes. But the same language does not fit the minute level where substance disappears into abstract energy and, in the famous words of Sir James Jeans, "the universe begins to look more like a great thought than like a great machine."[1] The conventional meanings of words are confusing in the newly discovered context of quantum physics.

At the other end of the spectrum, looking upward to our relationship with the world and the universe rather than downward to elementary particles, words are again a problem and for the same reason. The mega-scene—a new context that confuses everyday knowledge—is as difficult to capture with conventional language as the micro-scene. How shall we convey the ecological truth that people are not wholes but inseparable parts of Nature when that word has frequently been applied to the other-than-us, the wild, rough and crude, its meaning steeped in apathy, suspicion and even hatred? We can introduce new words and phrases—the Ecosphere, the world's sectoral ecosystems—in an attempt to convey a sense of the reality that encapsulates us, but such terms are unfamiliar. Their technical jargon jars, and people ask, "Are you not really talking about *Nature?*" An exploration of some of the ideas that cling conceptually to the word, and of what it should mean in the last decade of the twentieth century, is therefore worthwhile.

NATURE AS BRUTE

A major source of current environmental problems is in conceiving ourselves outside rather than within Nature, separate from rather than inextricably connected. Some political economists, lacking ecological insight, have emphasized the polarity between ourselves and the world by contrasting, in a pseudo-historical context, what man used to be before he civilized himself. The Hobbesian image of "the state of nature" imagines selfish savage individuals roaming the wild world like lone wolves, in constant fear of violent death, their lives solitary, poor, nasty, brutish and short. Out of this sorry condition, the social contract—a practical application of the Golden Rule—advanced humanity into the state of civility founded on mutual respect for the individual's rights to life, liberty and property. Doubtless the social contracts evident in society are real and important, but surely their values do not hinge on flights of fancy about prehistoric "states of nature" that we have somehow left behind.

"Nature, Mr. Alnut, is what we were put in the world to rise above," said Katherine Hepburn to Humphrey Bogart in *The African Queen.*[2] The equation, going back at least to St. Augustine, is sex = body = animalist = nature. Rationality looks down on such dross. Who in her right mind would want to go "back to nature" when it conjures up visions of being dragged by the hair into a cave? In a related way, many view "human nature" as the residue of selfishness and viciousness inherited from imagined ancestral "killer apes." Robert Ardrey is an enthusiastic proponent: "Man is a predator whose natural instinct is to kill with a weapon . . . . We are children of Cain."[3]

Nature is the label given to the unpleasant aspects of an imagined "way we were," convicted by association with grim fairy tales about how our ancestors lived and what was uppermost in their savage bosoms, by stories about moronic Flintstones whose thinly disguised suburban descendants we are supposed to be. Nature red in tooth and claw is the popular scientific version that reads competition into every biological activity.

According to this hypothesis popularized by Adam Smith, Malthus and Darwin, nature is a cruel arena of eat and be eaten, predator and prey, parasite and host, where only the strong and wily survive. In Woody Allen's words, "Nature to me is just a big restaurant." The idea is gratuitously perpetuated in TV Nature films that focus on predation—where the action is—while providing facile explanations of everything animals and plants do in terms of competitive advantage. Interpreted as Nature's norm and adopted as Social Darwinism, such one-sided theory invents a cruel world. It neglects the widespread co-operation and mutualism that bind species together in their societies and ecosystems. Omitted is the idea that Nature is the substantial home, the Home Place, within which we and other organisms have always lived, in the past as now, a Nature ever-changing and evolving, not necessarily matching our desires but nevertheless supportive, or we would not be here.

NATURE AS BITCH

Francis Bacon is usually credited with the most forthright description of Nature as capricious and bitchy maternal parent—the images are always female—who wickedly conceals or holds back what rightfully belongs to her needy and deserving children. He counselled the use of force, cajolery and deception to find her secrets or force them from her for the good of humanity.[4]

Caroline Merchant has explored the roots of nature-as-bitch, tracing the idea of Nature from Greek concepts of the cosmos as an intelligent organism to medieval ideas of an organismic world. As late as the Renaissance, belief was strong in the "affinity of nature" as a female world soul that bound together all things through mutual attraction or love. In the sixteenth and seventeenth centuries, the emergence of a mechanistic science and discoveries that revealed what seemed to be an unruly rather than a perfectly designed universe, eroded the old mother images and gradually substituted those of the witch. Merchant points out the strong sexual imagery that marked the initiation of modern science as men set about taming and subduing a disorderly Nature.[5] William James thought that attacking nature might be the moral equivalent of war, and modern statements of this spurious theme still recur:

> Once the world has been purged of ghosts or spirits, it reveals to us that the critical problem is scarcity. Nature is a step-mother who has left us unprovided for. But this means we need have no gratitude. When we revered

> nature we were poor . . . . But if, instead of fighting one another, we band together and make war on our step-mother, who keeps her riches from us, we can at the same time provide for ourselves and end our strife. The conquest of nature which is made possible by the insight of science and by the power it produces, is the key to the political.[6]

"When we revered Nature we were poor." The road to riches is obviously to switch from reverence to domination and control—a justification for the continued enormous expenditures of wealth on power-pursuing science/technology.

Here again we see how language and its usage has burdened the word *Nature* with a derogatory meaning that sets it apart from us. The tactic is similar to wartime propaganda that inflates our virtue while dehumanizing the enemy in order to kill him in good conscience. When Mother Nature is conceived as harlot or witch, we need feel no sympathy for her. We are not flesh of her flesh, and we owe nothing to her. We are important and she is not, a perilous falsehood.

NATURE AND THE CLASS STRUGGLE

Early in the 1970s, faculty members at the University of Saskatchewan organized an inter-disciplinary class called "Man and the Biosphere." In retrospect it should have been "People in the Ecosphere," but feminism was not yet in full flower, and biologists were first to put a name on the planet's skin. One evening a sociologist lectured to the class, undertaking to explain the sudden interest by the public in the environment. Observe, he said, that the decade of the 1960s was a period of strife, of social upheaval, of an unpopular Vietnam war, of student riots and labour unrest, the poor against the rich. Society cannot long endure that kind of unhappy disruption.

The interest in environment, quoth he, puffing on his pipe, draws everyone together in common cause, students and faculty, the rich and the poor, healing the wounds, knitting up the ravelled sleeve of social care. The dissention of the '60s gives way to the unity of the '70s, discord yields to harmony under whatever banner is handy, and "War on Pollution" happened to be in the right place at the right time. The pendulum swings from strife to harmony and back again, and so the world wags.

According to this thesis, still popular in the political arena, public interest in environmental matters is a brief respite from the serious business of bettering the social system. We struggle with thorny problems in the economic and political domains until pain and despair exceed a critical threshold and then, gratefully, we turn attention to trees, flowers and whooping cranes. Thus Nature is merely a diversion and a dangerous one at that. For it distracts attention from the real and important problems of this world: justice and equality for people, employment and paycheques for the labour force.

Some social activists, convinced that Nature has replaced religion as the opiate of the people,[7] suspect more than a diversion; they smell a conspiracy. Concern for Nature, they argue, is class-based and most alive among the bourgeoisie—people satisfied with the *status quo*—especially those who look askance at radical social change. How much better for them if public attention is skillfully turned to non-controversial things, to protection of natural areas, to protection of wildlife with the Royals leading the way. How much better to have everyone worrying about Exxon's oil spills rather than asking questions about Exxon's profits and the beneficiaries of the oil industry.

Such suspicion is buttressed by reference to interests of the proletariat who by and large, in city or in rural areas, show little concern for protecting Nature. The worries of the poor are sharply focused on keeping body and soul together day by day. They do not lie awake nights fretting about the salvation of the Swift Fox and the White-flowered Prairie Parsley. The words of an old-timer at the Churchill River hearings in northern Saskatchewan express their sentiments perfectly: "Fog the bald eagles; what about jobs?"

Admittedly "save the environment" slogans can be diversionary tactic for some individuals and nations who "have it made" and who want to preserve the socio-economic *status quo* against the claims of the have-nots. Such selfish dishonesty does not negate the fact that an environmental crisis is upon us, a crisis demanding that we take Nature seriously as the supportive surrounding matrix. Everyone is involved, and the issue should not be trivialized by classifying it as just another ideological facet of the class struggle.

An enlightened attitude to Nature is neither diversion nor conspiracy. What needs to be done affectionately for the Ecosphere will also illuminate the path we must follow to solve social problems, not only to even the odds between the poor and the rich, between the developing and the industrial nations, but also between disadvantaged future generations and this profligate one.

NATURE AS HEATHEN GOD

Particularly difficult for people brought up in the Judaeo-Christian tradition is escape from the idea that divinity has departed from the Earth, from Nature. For we have been taught that Pan and Baal and other nature gods are the adversaries of Yahweh, God of justice; that the locus of God's presence is in human industry, not in the world and least of all in wilderness. Nature gods are taboo, associated with paganism, animism, pantheism and other so-called "primitive" religions. We should remember that the root of primitive is *primus,* first, and consider that the first may possess as much insight as the last.

George Melnyk called for radical regionalism in western Canada based on the idea of a Métis culture, blending the best from immigrant societies and the indigenous Native societies.[8] Some find the idea surprising, for colonial thoughts are still strong among us. Did we not bring civilization to this empty land? Did we not bring the science/technology that converted soil and water from useless to productive, making two blades of grass topped with wheat kernels grow where only one small sterile blade grew before? How could a half-breed culture improve on what the new immigrants brought with them so proudly from Europe and Asia? What does the Native Indian side have to offer?

The answer is *respect*. The Métis culture that flourished briefly in Manitoba and then in Saskatchewan in the 1800s was a partnership, a synthesis, an adaptation forging equality between the native and the immigrant cultures. It remains a metaphor of the way we ought to live, not only with one another but with the land, with the Indian consciousness of its organic aliveness and its spirituality.[9] Among other things, the half-breed culture could teach us again that the Earth is sacred, that divinity has never left it and that we do wrong when we heedlessly tear it apart and destroy it.

The genius of Jewish theologians was to take God out of physical place and put him in the heavens, in the whirlwind and the lightning and, eventually, in the airiest place of all: in history. God is transcendent, above and beyond the reality we know, and this is necessary for a monotheistic faith. Yet such a theology carries a heavy ecological price. For if God's kingdom is not of this world, if He is not immanent in the grasslands, groves and waterfalls, then their sacredness evaporates. And with us, if nothing is sacred, nothing is safe. A popular Christian belief, pre-dating liberation theology though accented by it, is that the sacred realizes itself in two ways: in the church and in oppressed and exploited people.[10] Unfortunately this leaves the rest of the world fair game for the entrepreneur and developer.

Can Christianity support an ecocentric ethic? The jury is still out, uncertain whether the truth that the natural world is a primary revelation of the Divine can be matched with the theory that Christ, the God-Person, is the focal point of all history and the chief motive for all Christian living. Perhaps a positive answer lies in what Jesus proclaimed and represented rather than in what Paul preached. Charity need not stop with humanity but can be extended as well to other levels of reality. We are not confined in our ability to reach out and appreciate, and in fact we are as large as our love.[11]

NATURE AS HOME PLACE

Douglas Cardinal—Métis, shaman, architect—powerfully expresses his faith in organicism through the undulating lines and symbolic circles of his architecture. In his buildings, such as the new National Museum in Ottawa-Hull, the aboriginal consciousness comes alive, reminding us that the spirit moves in nature and that we are manifestations of the spirit. The native ethic that reconciles subsistence and coexistence is asking to be heard.

Caucasian immigrants introduced to the prairies the idea of Nature as separate from people. That view forced mutually exclusive choices. Will we choose jobs or wilderness? Will it be our welfare or animal welfare?—as if concern for Nature is necessarily opposed to humanitarian concerns. To argue for preservation of any parts of the wild world has been interpreted by some as a willingness to sacrifice people for horned owls, to bypass the needs of society for uncut forests—as if keeping the Earth-home in good repair is odious or simple minded, a sentimental rejection of reason, a preference for irrational instinct. In the extreme, concern for Nature is interpreted as a subversion of all that civilized people hold dear.

The old words used in the old manner frustrate our understanding. They continually turn attention away from our part-whole relationship with Nature. We humans are constituents of the Ecosphere, a truth that cannot be denied. The wildness of the wilderness is in us and we continue to need the wild world.[12] The ecological whole is inseparable from ourselves; we are coexistent and interdependent with it. Recognition of the fact does not denigrate ourselves, our society or our civility. Choices must still be made about how we will live our lives socially and individually, but from now on such choices must be made in an enlarged context, with enlightened understanding that whatever people do sets the tone of the whole.

New words, new language, new thoughts are emerging about ourselves in relation to the world and to the universe. Nature is neither brute nor bitch, neither diversion nor heathen god, though words and thinking can make her so. Nature is where we come from and where we belong in our Earthly existence. Nature is Home, with the responsibilities for care and affection and aesthetic concern that the word implies. To be at home means asking ourselves about our intentions of staying on, about care of the furnishings and their maintenance, about sympathy for the other occupants and their welfare—all matters with power to initiate a fundamental revolution in the practice of our arts and sciences, in time becoming our second nature, as we prepare to minister to the natural Home Place.

AGRICULTURE

LIVING WITH GRASS

From the valley of the Red River-of-the-North to the foothills of the Rocky Mountains, we Canadian-Americans occupy one of the world's great grassland regions, bounded by desert to the south and enclosed by various kinds of forest—coniferous and deciduous—on the west, north and east.

North American grasslands mark the transitional zone between a predictably arid desert climate and a predictably moist forest climate. Desert climates grow small thorny shrubs, forest climates grow large trees, and, in between, a sward of grasses and herbs dominate, surviving wind and drought, fire and grazing, by hiding their perennial parts underground. They draw back in the face of adversity, tough out the bad times and break out with renewed vigour when the good times roll. Their community is diverse, multicultural, consisting of many different groups able to accommodate in different ways to the unstable fluctuating environment. They demonstrate what it takes to survive and blossom in the rain-shadow of the Rockies. We should be so smart.

Prairie-dwellers never know what manna in the form of rain and snow will fall from the skies; we only know that our prairie home will sometimes be desert-dry, sometimes forest-moist and, on average, somewhere in between. We live in a zone where annual weather patterns may shift toward the desert or toward the forest, tracking the course of long climatic cycles that no one yet can fathom.

Understandably, early travellers and surveyors sent out from the east took back conflicting stories about the western plains. Sportsman-explorer Sir John Palliser, fond of blowing away grizzly bears, had hunted and travelled in the dry western-interior States a decade before his expedition in the late 1850s along the 49th parallel. He concluded that the southern Canadian grassland, later known as the "Palliser Triangle," was a northern extension of the Great

American Desert and too dry for farming. University of Toronto professor Henry Youle Hind agreed, based on what he saw while surveying along the Assiniboine and Saskatchewan Rivers in 1858. Whenever drought periods settle on the western plains, Palliser and Hind regain prophetic fame.

John Macoun, botanist extraordinaire, had a different vision. First hired by Sir Sandford Fleming to assess the potential of lands along the proposed Canadian Pacific Railway route, he preached the gospel of potential prosperity for all parts of the plains including the Palliser Triangle. In 1876 he told the federal committee on agriculture and colonization that, excepting areas covered with sand and gravel, "the greater part . . . is just as well suited for settlement as Ontario." Macoun is vindicated whenever precipitation arrives on time in the west.

Clearly, neither Palliser nor Macoun wrestled truth to the grassy ground. Weather experienced in the 1930s confirms Palliser's wisdom, while that of the 1970s supports Macoun. Even in the same year, Macoun may be right in the northern grasslands while Palliser is right in the south. Edmonton and Winnipeg are generally moister than Calgary and Regina; the black soil sub-zone on the Yellowhead CNR line is nearer the continuous forest and gets more moisture than the brown soil sub-zone on the Trans-Canada CPR line nearer the desert.

People in the west like to say that farming is a gamble. So it is, farm by farm, but over all the odds are known. We can expect droughts for protracted periods about three times a century. The Palliser Triangle was droughted out in the 1880s, the early 1920s and the 1930s. During the 1980s we've been teetering on the edge, experiencing some of the hottest years ever. Even in the black soil zone, dry years are to be expected. The 1961 drought not only withered the crops but also the aspen groves at the forest fringe.

We have spent one hundred years trying to beat the climate with technology, trying to make the grasslands behave like the forestlands of Europe and eastern North America from where, with high hopes, most of our ancestors came. Summer fallowing seemed to be the way to do it, advocated by Angus MacKay at Indian Head and W.R. Motherwell at Abernethy. Add together the moisture of two prairie years to make one forest year and a crop. It seemed a smart technique—until we found it destroys the soil's organic matter. Irrigation is another way to change the climate, but often it leads to soil salinization. How about biocides, to kill everything that soaks up water except the genetically engineered crop plants? The future of such a toxic technique is not difficult to guess. Maybe biotechnology will find a way to make grasslands behave like forestlands?

If we are to live in southern Alberta, Saskatchewan and Manitoba for another hundred years, we had better begin thinking what grasslands are and what, ecologically, they mean. Then ask, Can we devise sustainable agricultural systems attuned to periodic droughts and wet spells? What cooperative social structures and land use patterns would sustainable agro-ecosystems require? Should the Palliser Triangle be turned back to permanent pasture, to native grassland, and used only for grazing? Shall we soldier on to the bitter end, on sandy soil and on clay, putting our faith in technology and the power of prayer? Remember, if dry years did not come to the region where we live, it would not be grassland but forest. Therefore, to pray for rain in dry years is to beseech the Almighty for climatic change: "Please send weather fit for trees not grasses!"

How much better to pray, "Let me live in a forest," because that wish can be accommodated without upsetting the universe—simply by sending around a moving van.

GOALS FOR AGRICULTURE

We all know, at least in theory, that the goal of agriculture is to produce healthful food without diminishing the productive capacity of the land. The definition differs fundamentally from that offered by a former Minister of Agriculture. The goal, he said, is to feed the world's hungry and at the same time put an extra dollar in our pockets. Sustainability? It never entered his head.

People of the Western culture have worried little about Nature's sustained bounty, if they thought of it at all. When blight and famine struck the Irish in the mid-19th century due to potato monoculturing, they packed off to the new lands of North America. When overkill exhausted lamp-grade whale oil, the erstwhile whalers simply shifted to another oil patch. When the drought of the Dirty Thirties came to the southern parts of the prairie provinces, farmers headed north to the fertile parkland belt.

Our society has never had to face up to ecological reality because escape routes from man-made and natural disasters have always been found in new lands or in new technology. The times they are a-changing. Neither unexploited realms on land and sea nor the wizardry of technology can dependably sustain current food production, let alone supplement it for future needs. In the words of the wit, they stopped making land, but they're still making people, which explains why, relative to population, the globe gets steadily smaller. With no unpopulated shores and no more frontiers, boat people have nowhere to land, and refugees are a worldwide problem. At the same time, the other means of deliverance—technology—is revealing its dark side-effects of environmental pollution and destruction. Clearly a different approach to food security is required, an ecological approach.

AGRICULTURE AND CARE

Ecology means house/home-study, study of organisms in relation to the larger surrounding systems (homes) of which they are parts. Human ecology has to do with the living land, the organic systems of air-water-soil-organisms within which people live and to which they are related. If the species *Homo sapiens* is to survive, then human ecology should also mean the working out of right relationships between people and the land. From this perspective, the primary crop of agriculture, and indeed of all culture, is a caring people, for only they will treat the land as partner in a sustaining relationship.

Our ancestors were not stupid, and we find much wisdom in the language we inherit. Agri/culture means the cultivation of fields to produce crops. Within the words *culture* and *cultivation* is *cultus,* to care. Behind it, in turn, is the Sanskrit word *kwei* meaning to dwell with, as well as to care for. We are led back to an idea, deep in the language, that agriculture has to do with people dwelling on the land and caring for it.

To those who try to keep the culture in agriculture, to maintain the cultus or care in the people-land relationship, farming is a valued way of life and not just a business. This is the sentiment struggling for expression when people work to save the family farm, when they resist wholesale industrialization of their operations, when they try to reduce off-farm inputs, when they experiment with organic farming.

Such actions signal that all is not well on the land. The husbandman is being submerged by the attitudes and demands of the businessman. Maintaining any sense of interdependency with Earth, water and crops gets harder and harder. The caring relationship of farmer for farm is as endangered as the soil itself, both of them weakened and eroded.

SOCIETY'S GOALS FOR AGRICULTURE

To understand what has happened to the growing of food and how we came to the conventional and orthodox goal of today—greater production, higher yields—we need to look at our history. Conservation has not been its centre-piece.

Most would agree that we live in a materialistic society, born of an exploitive civilization. Those who came before us to North America set the pattern of "take from" rather than "care for." First fish and furs were creamed off for European markets followed by timber and minerals. When, in the late 1800s, a strong demand for wheat to feed the factory workers of industrial Europe created the incentives, preparations were quickly made for the settlement of the Canadian northwest. The herds of bison and elk were slaughtered, the native people hustled off to reserves, and surveyors arrived to inscribe a geometrical pattern on the landscape with no regard for natural contours. The road allowances and grid-road system that eventually followed defaced both hills and valleys.

No care was demonstrated in selecting the sites for towns; the railways simply dropped them on the map every dozen miles or so. One town, destined to become the capital of Saskatchewan, was located on a flat waterless glacial lake-bed, and the results of that particular bit of non-planning will plague Regina forever. The checkerboard allocation of land to corporations and individuals guaranteed isolation of farmstead from farmstead. Disdain for the prairie landscape and alienation from it was writ large on its face.

The driving policy from the start was production for export, building an industrial bias into the food system. From such beginnings, what hope for stewardship of the land? Nor were the social attitudes that came with the new settlers helpful. students of Canadian literature have pointed to the negative feelings brought to the western plains by immigrants from the gentler climates of Europe and eastern North America. Margaret Atwood finds one central theme in Canadian song and story: *survival.* Uncivilized and uncivilizable, is Wallace Stegner's appraisal of the southern plains.

Unlike the Amerindians who found the land bountiful, especially after acquiring horses from the Spaniards, occidental immigrants viewed the new land with a jaundiced eye, seeing in it an adversary, an enemy to be fought. The oft repeated saga of Western folklore and drama pictures the hardy farmer and his supportive wife triumphing over blizzards, hail, grasshoppers and dust storms. The blame for the misfortunes that followed invasion of a windy grassland region and the ploughing of its dry soil has been placed on Nature, not human stupidity. Suspicion of the natural order and faith in technology "to wrest a living from the land," provide a weak foundation for Earth-care.

CITY AND FARM

The cultural attitudes and thwarted expectations of Euro-Americans are not the only reasons for cavalier treatment of farmland. The roots of the problems that beset the western plains reach much farther back, perhaps to farming's beginnings when hunting, gathering, complex foraging and nomadic pasturing were supplanted by till agriculture that stirs the soil. The Romans named their goddess of agriculture Ceres from which came the word *cereal,* confirming that early agriculturists around the Mediterranean Sea grew rye, barley, millet. Such grains are readily stored and transported, allowing off-the-farm living in communities. Cereals provided the necessary food for *citification,* the literal meaning of civilization.

With surplus storable food, religious sites and meeting places for trade grew into hamlets, then towns and cities. Gradually a larger and larger percentage of populations became citified, urbane. The intellectual and cultural advantages were, however, purchased at a price. The ties between city people and rural landscapes were weakened as were sympathies for those who grow food. The good-humoured disdain of townspeople for country folk is indicated by a rich variety of put-down language: rube, bumpkin, clodhopper, hick, peasant, rustic, heathen, pagan, savage and even "farmer!" The latter finds only one phrase to slang back: "city slicker."

Cities are not primary producers; their industry concentrates on secondary and tertiary "value-added" products. Cities generate wealth by processing edibles, timber, metals and other raw materials whose sources lie in the hinterland. Rich ecosystems live off poor ecosystems, and cities, being rich, draw in a vast range of resources, not the least of which is food. Cities *must* have the produce of the countryside, but their presumed rights to it—the rights of the civilized—are not balanced by equal responsibilities for maintaining the resource base: the soils, the rural infrastructure. Yet no matter how irresponsible cities may be, they have the numerical, political and economic clout to determine, to a large degree, what happens out on the land.

In the modern city-agriculture system, the urban majority sets the goals and the methods to achieve them for the small percentage of the population on the land. Predictably from the cities—where live the agribusiness experts, the farm-credit bankers, the Pool and Wheat Board management, the university researchers, the politicians, the agricultural bureaucracy and all those consumers determined not to spend more than 18 per cent of their disposable income on food—the cry goes up for greater productivity, higher yields, more efficiency, more specialization, speedy adoption of new technology, bigger machines, more computerized management.

Look here, we've worked it out for you in simple figures, the city voices say. You should be producing 36 million tonnes of grain for export by 1990, or maybe 40 million according to Cargill. A few changes will be needed, such as getting rid of small and inefficient farms by consolidation. Saskatchewan's 71 thousand farms in 1971 were down to 60 thousand by 1988, so 40 thousand or 30 thousand by the year 2000 should streamline industrial operations. That some of you will be driven to the wall is an unfortunate fact, but this is the logic of economics and the market; sentiment has no place in this business.

Most everyone views the ruination of the rural fabric—the depopulation of the countryside, the disappearance of small farms, the death of small communities—with fatalistic calm and stoic equanimity. It's inevitable, today's market-sages say. Enough of sentimental nostalgia! Read Oliver Goldsmith's pastoral "The Deserted Village" bemoaning the same trend, back in 1770 for godsake! There's nothing new about farm consolidation and farmers being forced off the land; it's been going on since the beginning of time. The competitive system works its will, and naught can be done about it.

But let the centre of a Western city start to wilt and the fight is on to beg, borrow and steal whatever is necessary to breathe life once again into the blighted area. Urban renewal plans pop up like weeds through broken pavement, and somehow funds are found to redesign, reconstruct, redevelop and get population and businesses back together again. No passive acceptance of the laws of the marketplace there! No ritual chanting about economic necessity! And if it is a question of 40 million dollars for a new arena or football field, no problem! City ills are real and important; rural ills are suspect.

Obviously something other than market-place economics is at work when cities rally to rejuvenate their dying centres or to subsidize affordable housing. And something other than the serene impartiality of the competitive market system lies behind failure to re-weave the rural fabric. Subsidies may be handed out farm by farm because that kind of money flows back to the city, but rural communities die by the "laws" of economics.

Economics is a human enterprise based on human values. It is an invention, in the Western world designed to serve material betterment in accordance with ideas of personal liberty and individual aggrandizement, and it is fashioned to aid those with access to capital better than those without it. As an artifact, the economic system is never an impartial cause of any happenings, for it expresses the designers' ideas and intentions. Nevertheless, "economic laws" provide an excellent cover for the preferences of those in power and excuses for unethical treatment of the powerless. The city preference, implicit in powerful business interests, is to convert agriculture to factory-farming. Governments back the preference, then throw up their hands in horror at the trend.

Rural depopulation goes hand in hand with decreasing rural political power, and legislative policies favour urban over rural dwellers because political power is increasingly concentrated in cities. Agriculture is encouraged on its industrial path because, by and large, city people consciously or unconsciously perceive an advantage in it going that way. The priorities of urban society are cheap food, maximum exportable surpluses and sales to farmers of fossil fuels, machinery and chemicals. Who in the city can make a buck out of small self-sufficient farms or out of small self-sufficient hamlets or out of the discovery of cultural rather than chemical methods to control weeds or out of *reduced* production so as to rebuild the soil? Conservation farming sounds good, but if it means lower inputs and lower yields, it is not on the city's wish list. Conservation has a stagnating effect on the city economy.

This is not a "damn the city" tirade but an attempt to show that agriculture's problems are intimately tied to city attitudes. Because urbanization is predicted to continue in all countries, because more and more of us will live by choice in cities, we must recognize our direct and indirect influences on land use and agriculture. Can we, and will we, from our asphalt-and-glass environments support policies and programs of sustainability out on the farmland? It can be done. City people have shown they will support preservation of wilderness, an idea far removed from the urban scene, so why not strict conservation practices for agricultural soils as well as for forests and fisheries? The only lack is a consensus on the importance of farming well, of farming ecologically. The cost will be high but *this* cost is worth it.

ECOLOGICAL AGRICULTURE

Agriculture is ecological insofar as it is true to its original meaning, incorporating a spirit of care for the land that guarantees a sustained relationship between it and the user. Set against ecological agriculture in a variety of ways is today's orthodox farming, increasingly large in scale, mechanized and industrialized, a city-guided pursuit with goals at odds with care of the land. The trend is dangerous, and it is the trend—rather than those farmers who are obliged to go with the flow in order to survive—which merits the strongest criticism.

Most ecological axioms are, with a little thought, self-evident. The most important is that people exist *in* Nature (now denatured to "environment") as part of creation, sharing with myriad other organisms the fruitful soil, the water, the air. Looking out and around, especially in spring and summer, the planet is life-filled, alive.

Just as we are parts of the Earth, the Earth is part of us. The composition of our flesh is related to the organic matter of soils; our bones are another form of the limestone mountains; the sea and our blood are consanguineous. The world, said the philosopher, is your body, with rights at least equivalent to your own. The ethical imperative drawn from ecological understanding is coexistence, not exploitation. Put another way, maintenance of the Earth is humanity's first priority, which means protecting the capital and living off the interest.

Ecological rules of relatedness emphasize the dependence of humanity on other animals and plants and the importance of cycles by which nutrients pass in repeating cycles between plants, animals, soil, water and air. Natural ecosystems that have been self-sustaining for a long time tend toward a diversity of life forms. Variety is the spice of life.

Carried over into agriculture, ecological concepts place the farmer, the horticulturist, the gardener, as overseer of an interrelated web of life whose organic bank of capital is the soil. The system will go on and on as long as the soil-capital is not degraded by taking more from it than is restored. Sustained agriculture means maintenance first—attending to the health of soil, water, plants and animals—ahead of attention to yield and production. Schumacher's rules for land use are profoundly ecological: guard the health, beauty and permanency of the land, and don't worry about sustained productivity.[1] It is a derived benefit, a secondary quantitative benefit, that will look after itself if the quality of land is preserved. The primary accent on health is well-placed because the health of humans (and their beauty and permanency) cannot be separated from the health of Nature, a point that toxicology and the study of allergies increasingly makes clear.

PRODUCTION OR SUSTENTION

The central question is whether mainstream agriculture on the industrial model can or will take an interest in *maintenance* and *sustainable production* when its tunnel vision is on *maximum yield and increasing production*. With its priorities wrong, can it do anything but run down the soil? Can it support regenerative agriculture, or does it necessarily force degenerative farming?

The record in Canada so far is not good: loss of the soil's organic matter along with soil erosion, compaction, toxification and salinization. If such unrecorded costs of production—the price Nature's bank has to pay for an exploitive system—were included with costs of agribiz inputs, present crop prices would probably not cover the total.

True, these problems are now drawing attention. Research is focused on such techniques as minimum tillage to slow the rate of organic matter oxidation, stubble cropping to protect the soil surface, longer rotations with incorporation of legumes to add nitrogen to the soil, snow management to make up for the extra moisture that the old fallow system used to save, a greater variety of crops—pulses, oil seeds, cereals—to introduce more variety and therefore a degree of stability into farming systems, integrated control of insects and weeds with less dependence on broad-spectrum biocides. These and other conservation methods are a welcome sign of awareness that past techniques were damaging.

The purposes driving the system, though, are unchanged. The methods may be getting better, more conserving, more efficient, but the direction is steadfastly toward the idiot's goal of growth, toward higher production, higher yields, greater profits, towards more and yet more. Few are willing to face martyrdom by voicing the thought that maybe, just maybe, *good agricultural practice requires a decrease in production, not an increase.* As long as city hard-heads and their country converts call the shots, that possibility will not be entertained.

A PHILOSOPHY FOR ECOLOGICAL AGRICULTURE

Goals for an ecological, sustainable agriculture can be stated more convincingly once basic ideas of what is important are clarified. Unless values are examined and fundamental assumption questioned, we only skate around on the surface of vital matters. Without deep digging into a faulty system to discover the source of the breakdown, we condemn ourselves to be merely tinkerers and tire-kickers.

The machine analogy suggested itself without any effort on my part, and it signals one immediate danger. In our society, we tend to think in terms of machines and mechanisms. (The heart is only a pump, the brain just a computer). We have been taught by leading philosophers and scientists that the universe is a great mechanism: spiritless, purposeless, dead. Viewed as a machine, nothing can be wrong with studying it in ways that will give us control and domination over it. If it is dead, and only we are alive, then obviously it is an entirely different and inferior kind of being. Whatever we do to it for our own enrichment is perfectly all right.

This way of thinking has been extended to land, air, water and to other organisms. They too, we have convinced ourselves, are spiritless mechanisms. Destruction of natural ecosystems and all that is in them has, until recently, created little furor. Torture of animals to test cosmetics or to seek cures for human diseases and disorders—many of them induced by our own foolishness—is still acceptable to the majority.

The reigning philosophy about the nature of the universe and the Earth, known as scientific materialism, is driving humanity toward extinction. The world's slow destruction is proof of the wrongness of our central belief in a dead mechanistic world. Set over against this erroneous faith is an ecological philosophy that does not confine the values of life and consciousness to people alone but to the universe. Somehow the Earth has been peopled from star dust, water, air, sunlight. Life is a property of all these things brought together in ecological systems, of which the largest is the global Ecosphere. We live within this magical, vibrant ecosystem. It is not dead. How can it be when it infuses us with life?

The organic and ecological view of Nature, as opposed to the mechanistic view, is religious in the true meaning of *religio,* to bind together, to make whole or holy. If science is to be constructive and future-oriented, it needs such an organic framework, for otherwise it has a dissolving influence; it breaks things up, reduces, analyzes and anatomizes. Science teamed with materialism is dangerously destructive; hence today's need for the countervailing philosophy that confers values on the wider-than-human system, that is aware of the greater whole, that heals because it is organic.

An appreciation of the two different worldviews aids understanding of contemporary problems. Is the rural landscape, the farm, the soil, only a mechanism, a material machine for making food, as such phrases as *agrifood industry* and *agribusiness* suggest? If it is a machine, we can make it better because, although we cannot make life, we *can* make machines. That is our forte. So we set about levelling the land, grinding up the soil, reconstituting it with mixtures of chemicals, spraying it with poisons, aiming to make many blades of grass grow where one grew before. The insidious thing about the mechanistic approach is that *it works,* at least for awhile. Short-term success seems to confirm that Nature is nothing but a machine. But sooner or later the inevitable anti-life side-effects begin to appear.

Observe that machines lack spirit, and if all Nature is confirmed by our manipulative approach as nothing but dead matter, then the spirit goes out of us too. All elevated purposes of life disappear and naked selfishness fills the vacuum, operating with terrible effects at the level of the individual, the society, the nation. The superpowers say to each other, "If you threaten our rights to consume more and more of the world's resources, we will destroy you and the planet, fifty times over." Such insanity is only possible in a world conceived to be nothing but a machine.

If the universe is creatively alive, then purpose is restored. Those things around us that we have been taught to see as unrelated pieces of inorganic and organic matter are in reality the misidentified parts of a greater whole. Soil, water, land, air, are alive; they are the tissues of the Earth that sustain us in symbiotic relationships. Nature is not a machine, and no actions on our part are ethically neutral. Humanity is participating in a great creative adventure, a search for harmony to which our agriculture, like all other important human pursuits, must be tuned. Climatic change for the worse, erosion of productive land by wind and water, loadings of toxic chemicals in air, water and soil, public distress about the unhealthiness of food—all these separately or together may force modifications of farming methods. Solving such problems is the only way to ensure that agriculture continues in western Canada, as in the rest of the world. Efficient techniques on the old high production track will prolong farming for awhile, but efficiency—getting more for our money—is a trap and not the answer. Ecological agriculture is the answer, its goal to sustain the land's health, beauty, permanency and production forever. It is waiting in the wings, to be called on stage when the worth of the living world also makes its debut.

Agriculture and the food system, along with our attitudes to both, are parts of the greater culture of values and beliefs to which we all subscribe in cities and towns as well as in the country. When that culture absorbs and takes as its own the organic relatedness of people and land, then the care that seems natural between people will be also be restored to the fields—and that will mark a return to the original goal of agriculture and, through it, the redemption of all culture.

TOO LATE FOR EDEN?

ECOLOGICAL AND SOCIAL ISSUES IN AGRICULTURE

If the old Genesis story is to be believed, agriculture is the punishment for our sins. In the beginning, Adam and Eve were happy fruit-gatherers in the Garden of Eden, foraging for their meals, dining from the trees, practising neither horticulture nor agriculture.

As summer ended, they ate the forbidden fruit, lost their innocence and incurred the wrath of the Almighty who expelled them from Eden and condemned them, in the words of the Bible, to "till the ground." The penalty for transgressions is a life of ploughing the soil. Crime does not pay, and neither does farming.

The myth ties agriculture to mistaken choices. Perhaps the original sin was exploiting the soil instead of cropping—in the manner of gatherers and hunters—the seeds and fruits that the soil produces naturally. Live on the annual interest and leave the land's capital alone. Carefree foragers—women and men as equals in the diverse wilderness garden—may be closer to paradise than the worried farmer and his wife isolated in a monocultural sea of wheat or canola.

The past has its lessons for us, lessons that can help to shape a better future. Just as the Garden of Eden story casts tilling the soil in a questioning light, so do other aspects of the pre-agricultural world and of the human societies that occupied it. A historical perspective helps us understand what we are doing with food that used to be gathered and directly eaten for nutritional purposes but now is cash-cropped for free trade.

AGRICULTURE SIMPLIFIES THE DIVERSE OLD WORLD

Most natural ecosystems are organically diverse, supporting a great variety of plants and animals, fungi and bacteria. This is Nature's norm: a multitude of life forms coexisting as functional parts of the Earth's vital surface. Food-as-commodity production rejects this ancient wisdom.

Agriculture chooses a small number of organisms that can be controlled for its purposes. It exterminates the rest to make room for its selections and to remove competitors. In this way, agricultural societies simplify the ecological systems around them, tying human dependency to a few kinds of plant and animal crops. For several thousands years we have been cutting ourselves off from Nature's diversity and isolating ourselves with a smattering of placid domesticated organisms, thereby cramping the imaginative capacities that variety in environment enlivens.

Furthermore, agriculturally simplified systems are more chancy and unstable than natural ones that are less fragile and more resilient. A monocultured crop of wheat is precariously sensitive to the many natural hazards of wind storms, drought, locusts and pathogens, while in the same environment, native grassland—a polyculture of hundreds of different species—survives and grows year after year, come hail or high water. Some of the native grasses grow best when the weather is cool, others when it is hot; some herbs flourish when the year is dry, others when it is wet. No single cereal or oilseed crop displays that kind of versatility.

By harnessing the human psyche to the ups and downs of simplified ecosystems, to the boom-and-bust economy of a few selected crops, fear of material insecurity has been deeply implanted in our culture. Here is one source of the obsession with the accumulation of material wealth, a means of allaying anxiety and demonstrating personal worth. It may also be the origin of our love/hate relationship with Nature, who in agricultural societies seems to play the role of Capricious Mother, unpredictably giving or withholding her bounty.[1] Do the fears and insecurities of our society extend back to the dawning of agriculture? Eve's daughters may be inclined to say yes.

WOMEN AND AGRICULTURE

At the edge of the ancient city of Xian, in the heartland of China, excavation has revealed a neolithic village six or seven thousand years old. Interpretations of artifacts from the lower to upper levels indicate that the earliest society was matriarchal and communal; women played the leading roles. Later, as agricultural technology developed, social structures changed from female-oriented to male-dominated, and from communal to class-structured. In this part of the world, the development of agriculture apparently marked the fall of women from high status to low.

Perhaps the practice of agriculture requires more brawn than brains, compared to foraging at which women excel. Perhaps excess wealth in the form of stores of easily transportable grain was appropriated by the larger, stronger sex and then used as a power base to reorganize society on patriarchal lines. Whatever the cause, a shift from the nomadic gathering of nature's fruits to the sedentary tending of crops on the loess plateau of China seems to have brought with it the subservience of women.

Despite contributions of females to agriculture, including its invention, while the males were engaged in the more glamorous activities of fighting and the chase, the old names of the various branches—field husbandry, crop husbandry, animal husbandry—clearly show who has been in charge in Western agriculture. Wifery in rural communities has traditionally been confined to gardening and assistance with childbirth. Even today the family farm continues as a bastion of male dominance, perhaps the last because it is the strongest.

Women are therefore justified in viewing the culture of agriculture with a jaundiced eye as they react against ideas that it has historically reinforced. Men too have been victimized by the same social attitudes that compare unfavourably with those of their pre-agricultural ancestors. Cereal growing may be prerequisite for civilization, but it has not necessarily been conducive to a civil society. A reformed agriculture would be good for the spirits of both sexes, contributing also to their amicability.

On the personal level, paralleling the suspicious relationship with Mother Nature, Dorothy Dinnerstein traced psychological problems of both sexes to the Capricious-Mother image that many societies have foisted on women.[2] When men have gone off to attend to their interests, leaving women to assume the parenting role expected of them, the responsibility for satisfying all infant wants has settled squarely on mothers' shoulders. Because those wants can never be wholly gratified, mothers usually earn from their children a degree of mistrust and enmity. Thus the problematic relationships between ourselves and the wider world are paralleled in Dinnerstein's child-family problem. Like simple-crop cereal farming, single-parent child-rearing begets its appropriate psychological problems. The solution to both is discovery and adoption of more cooperative and ecologically sensitive ways of living and relating, so that we are rescued from forever "blaming mother" and "blaming Mother Nature."

IN SEARCH OF PLENTY—AND THEN SOME

The blame for the failure of agro-ecosystems to measure up to our exaggerated expectations is never-ending and will continue as long as agriculture pursues its impossible goal: complete food security. No matter what the weather, no matter what the population and no matter what the sustaining needs of the soil, we will coerce the land into ever-greater production of crops, into higher and higher yields, simplifying and controlling environment in the vain attempt to reduce risks to zero. Should we not instead learn from people who live or have lived the foraging life—as everyone's ancestors once did—taking with gratitude what Nature's ecosystems provide sustainably, not greedily pushing for evermore.

In Search of Plenty is the title of the centennial report of the Canadian Department of Agriculture, and the search goes on relentlessly. Each advance in technology seems to bring the goal of security and plenty almost within reach; each crop failure or reduction in yield due to uncontrolled elements in the environment shows the intransigence of Nature—the perverse mother withholding her milk. Another technological offensive is launched—new machines, new seeds, new poisons, new fertilizers—to squeeze out a further increment of production or at the very least to make sure that the level of production formerly achieved is maintained.

Biotechnology is the most recent assault weapon, expected to produce larger, faster growing, disease-and-insect resistant plants and animals. But more production for humans means a more rapid drain on Nature's components. If biotechnology succeeds, the deterioration of soil, water, air and other organisms will accelerate. Then amelioration measures—technological fixes such as doctoring soils and crops with chemicals—will add their unforseen impacts, further reducing resiliency and producing more unforseen side-effects in the world's ecosystems.

Committed though we are to agriculture, we need not continue forever on the hard path. Much can be learned from gatherers and hunters whose relationship to Eden, to the land, the Home Place, has always been less worrisome than ours. Foraging people live more in the present, taking as their model the lilies of the field as they grow. Foragers are aware of the implications of allowing population numbers to exceed the carrying capacity of the land. They are at home with environmental diversity and do not advocate simplification and forced growth. To the busy agriculturist, they seem carefree, careless and therefore imprudent—sins to a culture that tills the land, virtues for those who only gather Nature's surplus. Latecomers to North America are familiar with the accusation of improvidence levelled against native groups whose tradition was to live not as owners and exploiters of the land's assets, but as reapers of its annual dividends. Sympathetic study of their philosophy of life could inject much-needed wisdom into our unecologic culture and agriculture.

A SUSTAINABLE FOOD SYSTEM

The goal of continuous growth in food production within a finite world is unrealistic and will sooner or later lead to disaster. Pre-agricultural ecosystems were limited in productivity that was, however, stable and sustained at the level where natural nutrient loss and gain were in balance. Growth-oriented agri-ecosystems, force-fed with subsidies of fossil-fuel energy and chemicals, are unstable and ultimately unsustainable. They lack the wisdom of restraint.

Wedded to industrial and economic growth, the modern agricultural system serves a world population whose exponential growth it applauds. It uncritically accepts as its goal the continuous increase of farm yield. On this track, agricultural land is deteriorating everywhere in the world. To be sustainable over the long haul, production must level off at some steady constant that the land can maintain without forced growth. Population and per capita standard of living must be fitted to that constant.

The question of what the Earth—the soil and water and climate—can dependably produce directs our attention to natural ecosystems, to the ancient time-tested models, in order to design agro-ecosystems that are stable and resilient as well as productive. Wes and Dana Jackson at The Land Institute in Salina, Kansas, are experimenting with perennial polycultures of plants, selected from native Tall Grass prairie for their seed production, nitrogen fixation and natural herbicidal properties.[3] In their agro-ecosystems, tilling the land is minimized as is use of fossil fuels, fertilizers and biocides. Such experimental food systems are geared to running on solar energy as the natural non-polluting growth source. They are examples of *Low Input Agriculture,* as the food production system was until this century. They mark the beginning of the search for sustainable alternatives to the unsustainable high input, high-throughout agriculture on which agribusiness thrives at the expense of land and those who minister to it.

New directions in agriculture mean a changed philosophy, another turn of the progress spiral, recapturing a lost way of seeing the world and agriculture's place in it. The road to food security has to be rethought. The conventional approach, in the macho tradition of Francis Bacon, justifies the emphasis we give to knowledge, to science, as delivering command over nature. "Knowledge is power," said Bacon, "not mere argument or ornament, and I am labouring to lay the foundation . . . of utility and power."[4] For 350 years the idea that control and redesign of nature's processes can assure our security seemed right and good. Now we recognize that its assumptions are unecological, ignoring the dependencies of life on the ancient design of world ecosystems. Carried to its logical end, therefore, Baconian science in agriculture, as elsewhere, is destructive, and global deterioration is the proof that it no longer serves global security. A kinder, gentler, more nurturing approach is prescribed.

The new attitude, ecologically more realistic, defines useful knowledge as sympathetic and compassionate, contributing to co-operation rather than conflict with Earth's processes. In this direction lies security and the pleasurable experiences of at-homeness in the world. The salvation of agriculture, as of all human pursuits, is the relinquishing of power and control in favour of participation,[5] fitting agriculture to the ecological realities of the world and seeking a partnership with Nature as between women and men.

TWO KINDS OF SOLUTIONS FOR AGRICULTURE

Alarm about the deterioration of soils and productive capacity in western Canada has brought the question of agriculture's future into the public mind. Two kinds of solutions are championed: tinkering with the agricultural system and keeping it on the same high-production track, or transforming it radically so that it better fits the sustaining processes of the planet.

At present most people believe that agriculture's difficulties are minor; they propose to tinker with the system and improve it. They pin their hopes on more powerful technology to shape the mechanized greenhouse world beloved by Sci-Fi futurists. Others fear a disaster is unavoidable without profound attitudinal changes; they argue for a transformation of society, a changing of social values. Without a cultural shift, they say, the culture of the fields—agriculture—will not be reformed.

The tinkerers advocate conservation tillage, meaning such changes as snow management to increase the effectiveness of precipitation, the addition of phosphorous and nitrogen to the soil through formulated fertilizers, the greater use of rotations and leguminous nitrogen-fixing crops and substituting chemical poisoning for mechanical weeding—subtle violence for overt violence—to slow soil deterioration. Such techniques buy time and extend the life of conventional agriculture, but they are not the long-term solution. The system is still high-input farming, geared to maximum production, dependent on fossil fuels, wasteful of resources and of people—as the statistics on increasing farm size and rural depopulation show. On this track, agriculture continues as primarily a commercial system, an exploitive economic system, not an ecologically sensitive and sustainable people-Nature cooperative. It is male-managerial to the core, as indicated by the small female component involved in agricultural research, college teaching and direct responsibility for farming. Now and then it offers a conciliatory bouquet to Nature and to Nature-lovers in the form of a marsh undrained, a bit of native grassland unploughed, an aspen grove unburned, offended when instead of grateful praise it gets the deserved responses: What have you been up to now? Is this all you have to offer me? Small recompense for all the years of abuse!

The transformers advocate low-input farming, accenting the organic nature of healthy soil and good food. Taking natural ecosystems as models, they judge each agricultural technique according to whether it helps or hinders human participation in the world's renewing processes. The goal is harmonious with the nurturing female image: production of nutritious food at safe and sustainable levels, first for the home and region and only then for export to the continent and world—if indeed exports can ever be justified ecologically. To this end, agro-ecosystems must be designed to meet the expectation of the Ecosphere, the capabilities of the land, and not just those of the people in and on it. Mother Nature is accorded her due.

EVE OF A NEW EDEN

Eve dispensed food for thought in the Garden, and it is to the garden as a guiding idea that agriculture ought to return. The tree of knowledge is bearing a bumper crop of fruit in the late years of the twentieth century and much of it carries a warning note. Simplifying the world, overcrowding it, forcing growth beyond sustainable levels, attempting to make many blades of grass grow where a variety of flowering herbs grew before—these are recipes for calamity.

The agriculture we practice has not been patterned on a garden's beauty and diversity, the attributes of quality. Instead it is monotonous, aimed only at high yield, at quantity, reflecting a mindset close to insanity in that word's meaning of unhealthful and unholy. It is obstinately and unrepentantly crass and commercial, following the banner, "Farming is only a Business."

Turning food into money, like turning wheat into charcoal briquets and edible oil into diesel fuel, is not clever. The commoditization of food has disguised its real purpose which is to feed people and maintain their health. The commoditization of land conceals *its* real purpose which is to provide for all creatures—human and non-humans—a beautiful, permanent and productive home.

The commoditization of agriculture, of both land and food, confuses ethical priorities by setting production ahead of the land's needs and profitability ahead of human needs. Sane, humane agriculture should first meet the expectations of the land[6] and after that the expectations of the people who live from it.

Insofar as the past is interpreted ecologically, we can learn cultural and agricultural lessons from it, understanding how our societies fitted themselves to the cycles and flows and necessities of the Ecosphere's regional and local ecosystems. Such insights both warn against current practices and indicate where the new frontier lies. Unfortunately the North American dream was no more than the exploring and expropriating of another geographical frontier. It saw only a two-dimensional surface for the taking, land as property and the means to individual liberty—a vision insufficient to elevate the migrants' goals and aims above those of the nations from which they fled.

Now we know that the New Frontier, the true frontier, is circling the sun. Earth, the planetary home, is the discovery that lights up our lives. Home Place is the Eden where we have always been, even during the dark years when we thought we had lost it. Found again, it can lift us together—Adam and Eve, men and women—out of ourselves as we learn to cherish it and tend it as a garden, which is the meaning of paradise.

TRANSFORMING AGRICULTURE
VALUES, ETHICS, ECOLOGY

Before he began to write plays, George Bernard Shaw earned his living in London as a critic, first of music and then of drama. When asked how he could be a critic when he had composed neither music nor plays, he replied, "You don't have to be a hen to recognize a bad egg." My credentials are the same; I don't have a farm, but one need not be a farmer to know that something has gone bad out on the land. Daily the newspapers trumpet the "Crisis in Agriculture."

Thoughts about transforming agriculture are timely because traditional farming in western North America faces a new kind of disaster. Not the usual ones of drought, hail and grasshoppers nor the periodic market downswings of depressions and recessions, but more fundamentally a weakening of the foundation on which industrial production-oriented agriculture has been built: namely, the supplying of wheat for foreign markets assumed to be always there, always growing, demanding the "made in Canada" trademark, and able to pay in cash.

Just as people find it difficult to contemplate their own mortality, so few are prepared to contemplate the death of old-style agriculture-for-export in the prairie provinces. The pill is hard to swallow. We naturally suppose, having grown up with this particular farming system, that it will go on and on, forever.

Dan Morgan has written a capsule history of the world wheat trade, showing how various countries and regions had their day as exporters only to fade as world conditions changed.[1] Starting from about the year 1800, England and western Europe ran on coal and wheat, the one to fuel the machines of the Industrial Revolution and the other to fuel the workers of the machines. In such cities as London, Liverpool and Paris, the workers paid out about half their wages for bread.

For the first sixty years of the nineteenth century, the chief source of wheat for the European market was the territory around the Black

Sea captured by Russia from the Turks, with Odessa the grain-shipping port. Then the centre of the export trade shifted to the Danube River basin in what today is Romania. California's turn was next when, after 1874, it was discovered that the desert grass, wheat, thrived in the dry Sacramento and San Joaquin Valleys. California Number 1 was a standard variety traded by merchants and millers at the Liverpool Corn Trade Association, while back in the rural communities of the western USA, the familiar scenario of fighting the grain companies and railroads for decent prices was played out by farmers calling themselves the Patrons of Husbandry (the Granger Movement). History was to repeat itself fifty years later in Saskatchewan.

When the Suez Canal was opened in 1873, the comparative advantage in wheat trading shifted to India and after that to Chile, Argentina, Australia and then to the Great Plains of the USA and Canada, for Canada had benefitted by getting the short-season progenitors of Marquis wheat from India. By 1900 the California wheat trade was dead.

Along with the US producers of the Great Plains, Canada has dominated the wheat export market for fifty years. Not a bad run of fortune. Now we see the emergence of France as a surplus producer, and India again, and Mexico, and Saudi Arabia, and soon the USSR which is slowly going the way of intensive agriculture and pointing with pride to its "agro-industrial complex." Both the USA and the European Economic Community (EEC) have developed the capacity to overproduce grain at will.

The party's over, but in Canada we have become so used to dispersing *our* surpluses around the world (the original reason why the West was won) that we shout "Foul" when the EEC and the USA dump *their* surpluses abroad at fire-sale prices. Stealing our traditional markets! is the cry. Nonsense, replies the USA; you have had your share of shady dealings through the Canadian Wheat Board; we recognize no "traditional markets"—in a world devoted to freedom and free trade the rule is dog eat dog and every country for itself. If the USA wanted to argue the point from an historical viewpoint, it could point to the "traditional markets" stolen from California when *its* farmers went down the tube at the end of the last century.

This historic sketch is a reminder that global markets and suppliers change. Nowhere is it carved in stone that the purpose and manifest destiny of the Canadian western plains is to pour 20, 30 or 40 million tons of grain out of the country yearly, via unit trains and cargo ships, to serve a grateful world while benefitting agribiz and the national trade-balance sheet. Nor is it anywhere poured in cement that free trade and GATT bargaining will infallibly reveal, on the Canadian plains, a comparative advantage for cereal and oilseed production able to overcome all obstacles so that export agriculture on the industrial model can go on forever as Providence has ordained. The time is right to rethink what agriculture is all about, to begin considering new goals, possible alternatives to the old ways of doing things.

The system has not been working all that well anyway, especially from the standpoint of the farmers. The numbers of farms in Saskatchewan are half what they were forty years ago, less than sixty thousand and still declining. Rural communities—the life-support network for family farms—continue to weaken and disintegrate. About one third of the farming population is in danger of going under by the mid-1990s, especially if federal subsidies are shifted to projects that command more voter interest. If the subsidies that farmers themselves contribute by off-the-farm work were to dry up, the already rapid exodus from the land would accelerate.

Why then is society moving heaven and Earth to save and maintain an exploitive, industrial, export-based agricultural system that has poorly served a large sector of the farming population, while at the same time running down the soils, diminishing surface and subsurface water, destroying natural landscapes and decimating native fauna and flora?

The answers are to be found more in towns and cities than in the country. Farmers are the hired guns who willingly or unwillingly—for some have qualms of conscience—follow the directions of the urban majorities, the political leaders, the captains of industry.

Dennis Avery asked the question: Why does the USA keep on with massive over-production in agriculture? His answers apply equally to Canada. We do it, he said, because if we stopped pushing the export of agricultural produce we would:

1) renounce our export earnings,
2) worsen our balance of trade,
3) weaken our economic growth,
4) lose jobs in agribusiness and on farms.[2]

Losing jobs on farms is the least important reason. The truth is proclaimed in the newspaper headline, "Human Cost Forgotten in Farm Crisis."[3] Crises in agriculture are not identified by the *human* cost of farm families being squeezed off their farms but by *agribiz* suffering in the towns. A *real* agricultural crisis occurs when tractors, fertilizer and pesticides cannot be sold; when bankers see their loan business slackening; when the grain companies find themselves short of grain to store and market; when the CPR and CNR fall short of hauling to their full capacities.

At the top of Avery's list is export earnings—the favourable balance of foreign trade that agriculture provides so Canadians can buy fancy foreign products without going too deeply into debt: necessary things like armaments and other toys-for-boys that annually runs our defence budget over 11 billion dollars, all of it disappearing year after year without a constructive trace.

But the main reason for high production agriculture, whose goal is to squeeze more and more produce out of the land and dispose of it abroad by hook or by crook, is that high-production agriculture is high-throughput and high cash flow agriculture. It is capital-intensive and needs all kinds of city goods and services to make it work, thereby keeping the economy humming. As many have pointed out, the boast that each farmer on the land feeds seventy or eighty people in towns and cities is more realistically interpreted as seventy or eighty urban people making their living from the inputs and outputs of industrial farms.

A theoretical "best farm" under the present system is a big farm where annually the farmer gets a $250,000 loan from his bank then spends it in Saskatoon on machinery, fertilizer, pesticides, fuel, trucks, grain driers, augers, bins, silos, barns and housing. In the fall, he harvests his crop and sells it to the Wheat Board and Cargill for $270,000, receiving just enough to pay back his bank loan plus interest, and then he gets a job in town and his wife drives the school bus so that they can do it all over again next year. It's a living!

Just look at the cash flow on the ideal farm! The farmer has created half a million dollars worth of direct business, and much more indirectly. He may not have made a cent himself but he has laundered $520,000 for agribiz by running a high-cash flow, high-energy, high-turnover, high-production operation. And when the Ottawa and provincial governments come through with millions or billions in direct and indirect cash payments every year to keep the system idling in neutral, the farmers are encouraged to buy for the benefit of the needy in Winnipeg, Regina, Saskatoon, Calgary and Edmonton. Helping to keep the West alive is the high percentage of net income—estimated at more than 75 per cent in 1988—that farmers receive as subsidies from the public purse.

Note that the number of farmers on the land makes little difference to the way the system works, because borrowing from the banks and buying from agribiz increases according to farm size not farm numbers. In fact, the fewer the farms and the bigger, the better. For when they change from small to large, farms *necessarily* become capital intensive and totally dependent on high throughputs of goods and services. So agribiz prefers large holdings over small, and, in the USA, certain financial institutions in the 1980s set the minimum for farm loans at $250,000, forcing the consolidation of small farms into larger ones. "We can't afford to have people scattered out by making smaller loans," said one Vice-President who knows that bigger and fewer farms is better for business.[4]

Research in government institutions and at universities follows the lead of bigness, devoting most attention to projects that encourage and support high-yield, high cash flow agriculture on mega-farms. Why fight a worldwide trend?

In defence of it, have we not developed a highly productive, efficient, cheap-food system? And is it not needed to relieve global hunger? Close to one billion people, about one-sixth of the world's population, are undernourished, and with 90 million new mouths to feed each year—every eleven years another billion—should not Canada gear up to feed the starving world?

No, says the Report of the World Commission on Environment and Development chaired by the prime minister of Norway, Mrs. Brundtland. The recommended strategy for world food security is to shift agricultural production away from the industrial countries such as Canada to the developing countries where the real need exists.[5]

In the industrial countries, the Report says, agriculture's goal of production should be replaced by the goal of conservation, to head off the frightening environmental problems that are appearing worldwide: soil erosion, loss of nutrients, toxic residues from agricultural chemicals, depleting water supplies, shrinking natural areas. By making soil conservation a primary end, says the Commission, the industrial countries can develop less resource-intensive agricultural systems, sustaining their soils and renewable resources by attending to better land use, to water management, to forestry and to alternatives to chemical farming.

From its worldwide view, the Brundtland Commission recommends that we switch from production agriculture to conservation agriculture, substituting sustainable farming for exploitive farming. Canada's agricultural system can no longer be justified on the pseudo-ethical grounds that a hungry world must be fed out of the Canadian bin. This approach—not just by Canada but by other industrial countries as well—has proved destructive of the agricultural systems of Third World countries, as well as destructive at home of soil, water and wildlife.

Can the advice be followed? Circumstances are forcing at least its contemplation, and the hills are alive with the sounds of "sustainable development" and "conservation farming." Urged on by the missionary zeal of Senator Herb Sparrow, soil erosion in western Canada is receiving fresh attention from the Department of Agriculture's Prairie Farm Rehabilitation Program.

The question is, finally, whether we understand what is required to be conservationists after several thousand years of a Western value-system, an entrenched culture that has consistently advocated redesigning the world to suit short-sighted views of what is good for people. Headlines such as "Chemical Prices Blamed For Poor Soil Conservation" make one wonder. Will the millennium of conservation arrive when chemicals are dirt cheap? Are we running a market system or a system to provide people with healthy food?[6] Are we soil conservationists at heart or—as Hans Jenny, the Dean of North American soil scientists, suggests—only crop-production conservationists, intent on keeping yields and bank accounts high? We have all heard the refrain, "If it's not profitable it's not sustainable," a sad reflection on our priorities. True conservation means caring for things other than ourselves.

We are the heirs and victims of people-centred religions, of selfish ethical systems that attach values only to one species made in God's image, namely, us. We are the slaves of power-seeking science and technology whose aim is control, and of social, economic, educational and political institutions erected on misleading ideas of domination, coercion and competition with surrounding Nature, with plants and other animals, and even with our own species.

From stories in the media, film clips and articles, "people are led to believe that the Crisis in Agriculture is an isolated problem, a slight difficulty with one sector of an otherwise healthy socio-economic system. It is not. The Crisis in Agriculture is just one of many symptoms indicating that Western civilization has gone off in the wrong direction, chasing unrealistic goals defined by a narrow and myopic historical tradition. The Crisis is symptomatic of a culture out of tune with ecological realities, and the world's environmental ills are the proof of this fundamental disharmony.

The blame is shared by the two leading political ideologies. Capitalism is rapacious; it is a strategy of exploitation that puts capital before social needs. Socialism is rapacious; it exploits to provide for social needs ahead of capital. Of the two, socialism has the virtue of extending the circle of care beyond the selfish individual, at least turning our vision outward in the right direction. Concern for the collectivity of people may someday translate into concern for the greater community of Nature. Nevertheless, at the moment the sole interest of both is in people—capitalism to benefit those with capital, socialism to benefit the social body. Both express species selfishness which is itself the fundamental problem. Harsh censure should be directed to those supportive religions that foster humanity's self-centredness, especially the dominant sects that, against all reason, encourage human fecundity while paternalistically condemning the means of its restraint.

We need to back off and look at our species realistically. 6.2 billion strong, going for 10 billion, we exist courtesy of the thin shell of air/water/Earth that is the planet's skin—the Ecosphere (literally the home-sphere)—immersed in it, sustained by it, dependent on it, brought into being by it over four or five billion years, yet treating the Home Place as unimportant.

Problems with the environment, you say? We will look after them once we get the economy going again. Problems with farming as a major land use and alteration of the Earth's surface cover? Crank up the science and technology; we will find our way out with smarter tools and techniques. The message is power and control, putting one over on the Earth, making the Ecosphere serve us. It used to work when we were small in numbers and weak in technology. Nature then could absorb the punishment and shake off the insults. No more.

Preservation of the health of the world in which each one of us participates is the meaning of Conservation. *That* health is being sapped by a technology demanding too much, a technology that poisons as it takes, paying insufficient heed to recycling and restoration.

The crisis in agriculture—as in all human enterprises directed to high-energy growth, to extracting more and more from the planet's ecosystems—is part of the great environmental crisis that makes headlines every day. The necessary transformation of agriculture will come only as a consequence of a transformation of culture, when humanity recognizes the surpassing value of a world that transcends the species.

In mid-century Aldo Leopold preached the need for a land ethic.[7] Indeed we do need a land ethic, expanded into a world ethic. But exhortations will not bring it. People attach values only to those ideas and things that they deem truly important. Before fundamental changes come about in our institutions and in our ways of doing things, ideas have to change about *what on Earth is important.*

The scales will drop from our eyes when we understand ourselves as one species among many, the consciousness of the world because gifted with consciousness, sharing a marvellous, miraculous, creative sphere that somehow made us and which, once destroyed in part or in whole, we cannot recreate. We have to see that beyond ourselves, truly, a larger Being exists.

In this context, the outlines of a transformed agriculture can perhaps be seen. Because the habitability of the Earth depends on solar-powered processes and the resources they renew and replenish, we will have to tone down and eventually phase out the subsidized high-energy systems—dependent on non-renewable *unnatural* resources—such as those we have devised for industrial agriculture. Low-energy, low-throughput agriculture—one of whose manifestations is organic farming—must be the way of the future.

The implications of such an ecological agriculture for farm size, for numbers of people living on the land and looking after it, for supportive research, for decentralization of the social system, for the shaping of our cities and for world population size are profound.

PRAIRIE LAND AND PEOPLE

Each year in the Tall Grass Prairie at the Land Institute near Salina, Kansas, Wes and Dana Jackson sponsor a Spring Festival. The theme in 1988 *was "Health, Beauty, Permanence," an apt focus both for the land and for the communities it nurtures and a celebration of the trinity from which sustained productivity flows. Following are some thoughts from a talk given at the Festival.*

LAND AS CONTEXT

All societies and cultures exist dynamically within the Earth-space we call *land,* whether precariously or in ecological balance. Yet the idea that the Earth on which we walk, the air over it that we breathe, the water we drink and the food we eat are all parts of encompassing Land has somehow escaped our notice and admiration.

Everyone knows that what goes on *inside* a human body explains its functioning, its physiology, while the significance of what goes on *outside,* its ecology, is not so evident. Both, of course, are important. By looking both in/side and ex/ternally, in/specting and ex/amining, in/vestigating and ex/ploring, reality is uncovered. Thus, to understand things, we "look into" them and also "find out about" them, in both ways discerning whole-part and part-whole relationships.

Ecology is a reminder that context contributes one-half of understanding. Applying the ecological perspective to ourselves, by analogy with birds in the air and with fish in water, the Earthly milieu that sustains and recreates us is discovered, the Land-as-Source—much more than a resource. First was the land and from it came humanity, formed from its clay. Although we lay claim to the Land, buying and selling it, more realistically it can claim title to us.

A familiar poetic theme portrays people asserting brief possession of the Land that soon repossesses them. In Emerson's poem *Hamatreya,* the Earth sings:

They called me theirs
Who so controlled me,
Yet everyone
Wished to stay, and is gone.
How am I theirs
If they cannot hold me
But I hold them?[1]

In this and similar poems, the motif of human self-importance, of homocentrism, is asserted. The man-managers take charge with bravado, staking their claims to the planet's surface. The Land endures and waits, at length reclaiming the claim-stakers as its own, quietly affirming its encircling importance, asserting its grave truth.

THE LAND IS PRECIOUS

Schumacher called for philosophical and religious changes in how we treat the land, opining that *next to people* the land is our most precious resource.[2] Aldo Leopold disagreed. What we thought was our resource, our commodity, he said, is really the community to which we belong.[3] People are certainly precious—but next to the Land. Note that conferring primary importance on the Land, on the world that comprises all land and water, air and organisms, need not devalue its constituents—people included. Such recognition provides an elevated purpose for humankind: to be the conscious sense of the world and to treat it well. By such altruism we would by no means harm ourselves.

We cannot *mind* the world until the world is important in our minds as the surrounding inventive and sustaining sphere, a thing of intrinsic value and an end in itself. Maintaining its *Health, Beauty, Permanence* in perpetuity is perhaps the only creative, risk-free goal that humanity can set for itself, a goal, moreover, guaranteed to restore health and beauty to humanity.

Were Earth's ecosystems highly valued and protected, man's inhumanity to man would be reduced—especially in matters involving land as private property and the license for personal aggrandizement and exploitation that its possession allows. Mutilation of the world is a disfiguring of culture. How can one hope to be personally whole in a dismembered society, a land scalped, scraped, eroded, and poisoned? asked Wendell Berry.[4]

APPRECIATING THE GRASSLANDS

Love of the land is learned in the bioregion where each of us is born. I was brought up in the grasslands of western Canada and have spent most of my life here. The wide-open landscape, the big sky, the singing grass, the meadow lark's song, the wind-waves that roll through the grain fields, the indigo water of prairie ponds at spring breakup were imprinted on me at an early age. And so it seemed right, at the end of the Second World War, that I should enroll at the University of Nebraska in Lincoln to study grassland ecology with Dr. J.E. Weaver. He took me to a half square mile of native prairie near the city of Lincoln and in effect said, go to it, find out what you can about it—a marvellous introduction to the Tall Grass Prairie, and the ideal way to begin an education.

J.E. was a serious, life-is-real, life-is-earnest kind of person, some would say the model of a hard-headed scientist. Nevertheless in his book *North American Prairie,* he allowed a momentary exposure of his affections, writing of the grassland: "One glories in its beauty, its diversity, and the ever changing patterns of its floral arrangements. . . . One comes not only to know and understand the grasslands but also to delight in them and to love them." And on a nostalgic note, perceiving the imminent invasion by imp rovers of the grassland, he wrote: "The writer has been fortunate in living in that portion of the prairie that has resisted civilization longest."[5] Are there still some in Lincoln resisting civilization?

Resistance is more and more a duty. The immigrant civilization that came to the plains from eastern North America and Europe imposed on the grasslands a foreign technology, developed in quite different ecosystems. Over most of its ten thousand-year history, agriculture was practiced on naturally irrigated deltas and floodplains and, later, on deforested lands of ample rainfall as in Europe and eastern North America. Only in the last 125 years or so have the semi-arid grasslands of the world been tested. Dryland cropping in northeast China, for example, is only eighty years old, somewhat less than in western Canada.

No precedents exist anywhere in the world to indicate whether semi-arid agriculture can be sustained for much more than a century. A giant experiment of uncertain outcome is underway on the Interior Plains from Saskatchewan to Texas, to test under stress the resiliency of the prairie. Loss of organic matter, loss of soil structure, wind and water erosion, all are indicators that the stability of grasslands depends on a permanent grassy cover. The long-term prescription for sustainable use, for kindly use, may well have its pastoral side—shepherding flocks, propitiating the spirit of the primeval bison.

FORESTLAND HOPES IN A GRASSLAND CLIMATE

Even when the grass lands have been recognized as different from the forested lands where traditional agriculture developed, the hopes brought to it have been the same. Many have been the theories that discount drought as the reason for the absence of trees. And still today inhabitants of the Interior Plains have not come to terms with their semi-arid milieu, nor with the Corollary that sparse moisture is the norm to which agriculture must accommodate. The expectations of forest-dwellers continue to persist on the plains, a recurrent source of disappointment, frustration and despair. Instead of accepting and exploring what it means to be Plains People we continue to act like disappointed foreigners, finding much at fault with our Home Place.

Edward Ahenakew, an Amerindian of the northern grasslands, expressed his sense of himself as part of the Land in these words: "I am a Plains Cree," he said, "and on the prairies I can believe I am the centre of the world, my world. The land of the forest is a good land but when I am in the midst of it I feel small and of no account."[6] Until we accept our milieu as "the centre of the world," we will not treat it with care and respect. How much harm has been done and is being done to the grassland bioregion by those with unreasonable expectations of it, by those with no empathy for it because they came from the land of the forest, feeling themselves small and of no account in the strange surroundings of the grasslands.

PRESERVING THE PRAIRIE

One sign of respect for the prairie is the desire to preserve, and happily that desire is growing, though not quickly enough. The western plains are the centre of numerous species-in-jeopardy programs, of salvation strategies for black-footed ferrets, whooping cranes, kit foxes, trumpeter swans, burrowing owls, the prairie chicken, ferruginous hawks; and the reasons are not far to seek. The massive plough-down of the grasslands that continues, even when grain cannot be sold, has made this interior region one of the most intensively utilized agricultural areas in the world and one of the poorest in native landscape ecosystems.

Only fragments remain of the rich and varied True Prairie that once covered almost seven per cent of the conterminous United States—and that should pain us all. At the north end of the Tall Grass Prairie formation, in Manitoba, an inventory is underway to locate the last few remaining patches, and the suspicion grows that the largest is a thirty-acre block surrounded by houses within the city limits of Winnipeg. In Iowa the last major native grassland is reported to occupy no more than two hundred acres.

May Watts lamented that the tall-grass prairie survives only along field fencelines: half a mile long and eight feet deep to the root tips. Cornfields on one side of the fence and small grains on the other, she looked and saw Illinois as a one-foot-wide strip sandwiched between the alien vegetations of Central America and Europe.[7] Kansas deserves the palm for preserving the single largest True Prairie tract, the Konza Prairie Research Natural Area, saved from the plough by affectionate foresight—or by the obdurate soils of the Flint Hills. And of course other lovers of the prairie have preserved precious patches of the primordial landscapes, such as the native Wauhob Prairie near

Salina. Stephen Pyne tells us that these grasslands for ten thousand years or more were "conflagrated prairies" that in a sense are cultural artifacts, resulting from the use of fire as a technologic tool by the aboriginal inhabitants of the Interior Plains.[8] Some ask if present land use—till agriculture—is not just another technological innovation displacing the aboriginal one, ploughed lands in place of burned lands?

We must, I believe, defend or condemn technologies not only according to what they do constructively or destructively to the world, and to people, but also according to the attitudes that beget them and which, reflexively, they reinforce. Without romanticizing the life of the Amerindians, they—like foraging people everywhere—lived lightly on the Land. The evidence is strong that they shared a land-community ethic. Hence when they "conflagrated" the prairies, adopting a tool from the natural regime of prairie fires, it was not to overpower the Land but to help in its recreation, as a partner. The same re-creative goal is in evidence at the Land Institute where Dana and Wes Jackson are searching for ways to reform agriculture.

Despite the few glowing sparks of preservation and conservation that locally light the prairie landscape, a general gloom prevails. We have transformed rich ecosystems into poor, sacrificing quality for quantity, using technology to simplify, exploit and mine the Land. And we have done it mostly without qualms, following the scriptures' literal admonition to dominate the Earth. We are the elect, lords of creation. In the words of the Bard, "noble in reason, infinite in faculty, in action how like an angel, in apprehension how like a god, the beauty of the world, the paragon of animals," Shakespeare had the good sense to end Hamlet's recitation with a cautionary "And yet . . . ."[9]

And yet the message, any way you slice it, is species self-love. In a modern version of the old Greek fable, Narcissus took a shine to one of Zeus's women. Shaking an admonitory finger, Zeus warned him, "Just watch yourself!" *Homo sapiens* has taken its cue from Narcissus; its sole concern its own image. Where is the escape route to rescue us from the solitary confinement of a narcissistic existence, from the cramped, introspective cage that the human species has willed for itself?

On the authority of a number of seers, simple exposure to the Land, to the natural world—the earlier the better, and the wilder the better—can work miracles. Next to the Land, close to its creative powers, people too become precious.

THINKING OUR WAY OUT

Faced with today's mammoth environmental difficulties, people turn to science and scientists for solution to problems of hunger and food supply, problems of over-population and soil destruction, of industrial pollution and threats of nuclear war, of global change and deteriorating forests. Reliance for advice on those whose mission is problem solving, discovery of new knowledge, comes naturally in a society where the secular religion of science rides high. Apt to be overlooked is the fact that the ozone holes over the poles discovered by scientists and monitored by them on our behalf were created by an earlier generation of scientist/technologists playing around with unknown-to-nature chlorofluorocarbons.

Can such trouble-shooters double as guides? Might it be that their very mode of knowing—objective, analytic, manipulative—fragments and obscures the reality of wholeness? Is it better to look to the other solitude for leadership, to the scholars? Or are they part of the problem too?

Surely philosophers, artists and historians bear no ecological guilt! Surely the pursuit of goodness, beauty and truth is environmentally benign? Dead wrong. These comrades have also played their part in setting the course that has led to the present morass. Because they express the beliefs and values of the culture, extricating ourselves will surely require their reassessment of the old travel plans. The movers and shakers—the scientists, engineers and sodbusters of this world—are only half to blame for their errors. On society's behalf, they pursued goals that the scholars had endorsed as valuable and desirable.

The new perspective, Land First, is an ecological vision that *all* must share and propagate in the interests of health, beauty and permanency. The Prairie where we are today, the land of the grasses, needs its sympathetic philosophers, artists, historians and scientists to stir the imagination about what once was and what can be, with images that are ecologically realistic, based in the bioregion, true to the True Prairie, transcending traditional myths of heroic humanity fighting dust and drought, grasshoppers, blizzards and bad markets.

In our minds, we must find our way back to the lost place where we have always been—the globe, the Ecosphere, the Land, the Home Place—seeking a common vision and then a common covenant,[10] to understand, to respect, to love it.

TRADE AND TRAVEL

RESTRICTED TRADE

The goal that Ottawa pursues on our behalf, so we are told, is a prosperous and just Canada, and the only route to its achievement is expanded competitive commerce. Canada is a trading nation, and every expansion of foreign markets, every international agreement that encourages trade, helps Canada.

If you hear the credo often enough it begins to sound reasonable. Eventually repetition makes it an article of faith, the unassailable dogma. It must be true because economists repeat it endlessly. It is their religious chant, their mantra that quiets the rational mind and induces profitable inspiration, glowing visions of ever-expanding economic growth and affluence.

The questionable part of the creed concerns benefits over the long haul, for none can dispute the high payoff in the short term from trading resources for cash. Canadians lucked into a large chunk of Nature's bounty, and for several centuries they have enthusiastically busied themselves with its disposal.

In the beginning, the export of beaver pelts made a considerable number of non-natives rich, setting a standard for the exploitive skin-game that continues today. Next the white pine forests of the east were creamed off—providing masts for the Royal Navy—followed by exploitation of other easily accessible resources: lumber, pulp, cereals, oil seeds, petroleum, natural gas and all manner of minerals. Uranium is the latest in the long list of "goods" shipped out to a grateful if improvident global community.

Such trade has made Canada the planet's wealthiest Third World nation, its richest underdeveloped country. Canada is a super commodity-exporter, a "staples economy." Yet Canada is a pitiful provider of its own needs. Is Canada, perhaps, a sucker? Suppose that by some mischance this fine piece of real estate with its relatively small population were the only country left in the world, an island unto itself.

Suppose no hungry industrial giants existed, no nations clamouring for fiber and food, minerals and energy, no one to soak up Canadian exports.

Would we starve because we had no one with whom to trade? Would we be unable to clothe and shelter ourselves? Would we necessarily lack the amenities of civil living? Would Canadians, attending to their own welfare, using their own means, be fated forever to experience the "poverty of protectionism?"

The conventional answer is that we could certainly survive, but miserably. Foreign trade brings in more cheaply than we can produce whatever things we lack. Export of staples generates the wealth for buying goods and services. And so the conclusion: the way to increase the nation's wealth is by selling abroad, bringing in dollars that internal trade could not generate.

Notice that the argument is in terms of exchange values—wheat for television sets, pulp for celery in the winter—not in terms of the abilities of nations to satisfy their own requirements. Free traders do not argue that naturally rich nations like Canada are incapable of looking after their citizens' needs. Foreign trade's justification is cheap goods and services. When countries or blocks of countries reject this philosophy and opt for self-sufficiency, for example in the agricultural sector, cries of outrage ring out from those who have a comparative advantage and can produce cheap cereals, vegetables and wine.

If we ever adopt it, the national goal of sustainability rather than cheapness will put a different face on international trade. Suppose that Ottawa elevated, above wealth and prosperity, the idea of keeping Canada—its soils, forests, lakes, rivers and its population—in at least as good a condition as existed in the middle of this century, in 1950. That would require a certain amount of rebuilding, of restoration. Such an ecological goal would also dictate looking after the country as if it were an island, fully responsible for itself, without depriving other countries of the resources they need for similar reconstruction. The negative side of foreign trade would be exposed.

One immediate benefit would be forced interprovincial trade

and co-operation, in the interests of the country as a whole. Instead of every provincial premier building an energy empire with a sharp eye on piping gas, oil and kilowatts into the USA, Canada would have an energy policy worth the name as well as a country-wide energy delivery grid. With national access to renewable hydroelectric power and attention to conservation, the dangerous nuclear option could once and for all be laid to rest.

The big winner would be the Canadian environment, as well as our descendants. Foreign trade in the primary production sector, where Canada shines, is an unmitigated disaster environmentally, because it destroys and takes away without replacing—as witness the depletion of the nation's forests, fisheries and agricultural soils. Linking rates of exploitation to prices set by foreign demand, rather than to the natural processes of replenishment, guarantees that both forests and agricultural soils, as well as inshore and offshore waters, will intermittently be mined in phase with upward swings in international markets.

A sustained environment is by definition one that receives back in quantity and quality as much as is taken from it, and one that is kept free of destructive pollutants. Sustained forests should not be equated with artificial monocultures, made by slapping in plantations of fast-growing trees at two-metre spacing on ploughed soils after the old-growth is logged. Sustained forests are forestland ecosystems reconstituted in such away that they will develop into a likeness of what they were—complex and diverse in species—before logging removed the tree component. Similarly, sustained lakes and rivers will be restored in water quality and quantity by appropriate land uses in their drainage basins. Agricultural soils will be sustained by replacing the humus lost by tillage and the nutrients leaving the farm-gate in harvested crops.

Clearly such attention to conservation, to sustension, requires recycling. The replacement of what is extracted from productive ecosystems is most easily accomplished locally and regionally. It is almost impossible internationally and globally. The nutrients in prairie wheat shipped out to feed the world will never find their way back to the soils of the Canadian west. Export of resources means loss of their control. Foreign commerce kills recycling. The sustainable unit of the Ecosphere is the bioregion attending to its own needs.

In international trade, we bargain with "foreign" nations, which gives the transactions a character more abstract than those carried on at home. The ethical sense is dulled in dealing with far away places that have no apparent ecological connections to us. The exchange of questionable products is therefore encouraged. We may know that chlorinated pesticides and uranium are harmful, but the export of such stuff seems to pose no problem to us. Besides, if we don't do it someone else will. Such thinking has released on the world an array of dangerous products that internal trade, inside Canada, would not condone. To free up trade, to encourage it, to make it the centrepiece of government policy, is to invite ecological disaster. Trade *within* Canada where the health of ecosystems is accessible to political control is one thing; international trade that surrenders control is quite another.

The first goal of the citizens of any country should be to assure its sustainability. To that end, Canadians should begin thinking about *restricting* trade to what can safely be managed, given the goal of sustaining the national environment. Restricted trade, based on what the land and people can produce beyond their own needs in a conserving way, will also slow consumption here and abroad, a necessary step toward securing a livable Canada and a livable World.

In England in 1770, Oliver Goldsmith saw the ecological handwriting on the wall, though at the time, of course, he was unaware that migration to new lands, the unlocking of fossil fuels and the discovery of machine technology would postpone for several centuries the day of reckoning. His message that we ought to be developing our own local strengths, our own self-dependent powers, rather than staking everything on a desperate world-trade gamble is fitting today:

> Teach erring man to spurn the rage of gain;
> Teach him, that states of native strength possest,

Though very poor, may still be very blest.
That trade's proud empire hastes to swift decay,
As ocean sweeps the labored mole away;
While self-dependent power can time defy,
As rocks resist the billows and the sky.
(Lines from "The Deserted Village")

TRADING IN WATER
CURRENTS FOR EXCHANGE

Sell water to the US of A? Why not?! Every school-child knows Canada is prosperous because of natural resources, the inherited bounty of rich agricultural soils, forests, fish, minerals, coal, gas, oil and water. That these treasures were providentially placed ready at hand for the benefit of a deserving people is self-evident. So too is the corollary that failure to use them to the full, as rapidly as possible, is a squandering of wealth. Free-running undammed rivers, like wilderness forests and shut-in oil, are wasted.

The companion term to natural resources—raw materials—implies that their normal purpose and end is processed improvement. The raw and the crude were meant to be refined and developed. In the maturity of the industrial age, convention confers little of intrinsic value on resources beyond their potential to contribute to individual and public wealth.

Projection into the future of this way of thinking has called forth from the business and university communities an army of resource developers and experts trained to cost/benefit the value of everything in the optimizing search for the Best Deal. Encouraged by the pronouncements of free-trade bargainers, they are preparing to argue the advantages of selling beyond Canada's borders the fresh water that yearly flows over the plains and through the forests to drop at last, terminally and "uselessly," into the sea.

The alchemy of turning water into money has seized the imaginations of not a few provincial leaders, consultants and resource economists. It seems so easy, so natural. After all, Canada attained its status in the world and its high standard of living by the export of resources: first furs and timber under imperial colonialism and then, under economic colonialism, wheat, wood, pulp, minerals and energy. Why not add to the list of commodities the north's excess of water? How much less wasteful to channel it southward, reaping a fortune while earning the plaudits of a thirsty and improvident neighbour.

The stinger—and there is one—lies concealed in the word "resources" that uncritically lumps together things of vastly different value, while conferring on all a narrow economic meaning. Some important resources, of which water is one, are also key components of environment. Their unwise manipulation can erode environment's foundations and diminish its potential for supporting life.

Water is a different and more fundamental kind of resource than, for example, the trees that it nourishes. A person who grows trees with care should be able to produce an exportable surplus for many years, as long as the sun shines, the air is maintained unpolluted, the rate of soil restoration is commensurate with nutrients lost in the crop, and the waters continue to flow.

But she who diverts for export some percentage of the land's surface-water, exports environment itself, as surely as if she were redirecting rain clouds from southern Alberta across the forty-ninth parallel to irrigate Montana or redistributing to Alaska by satellite-mirrors a fixed fraction of the sun's radiation that lights and warms Ontario. It is comparable to trading away, by long-term contract, some proportion of the soils as they are slowly constituted on the prairies or packaging for sale (if it were possible) slices of fresh mountain air as excess to British Columbia's needs.

In short, the hewer of wood and the drawer of water are in vastly different businesses. They deal with "resources" of fundamentally different kinds. They should not be granted the same license to export.

Each river's regimen of flow is important to the form and structure of the landscape of which it is a part, as well as to the life in and beside it. To transfer water from one drainage basin to another enriches the receiving ecosystem while impoverishing the one from which the life-blood is diverted. The counter-current of money generated in the exchange cannot compensate the donor in kind and must always be judged a bogus bargain.

Nevertheless, all things considered, does it really matter whether water leaves Canada through engineered exits at the USA border or through the natural mouths of rivers on the shores of Hudson Bay, James Bay and the Pacific, Arctic and Atlantic Oceans? Yes, it does matter, not only to people in their land systems bordering and dependent on the rivers where they flow, but also to those oceanic ecosystems whose characteristics and productivity are linked to the discharge of fresh water with its sediments and nutrients in the off-shore zones. Water export will put both the terrestrial environment and the marine environment at risk.

Most importantly, plans to trade water as just another economic commodity express the dangerous tendency to rush technological solutions to all scarcity problems. Worldwide environmental trends indicate that humanity will survive only with a change in attitude that places high values on the planet just as it is constituted, an attitude that encourages people to accommodate to the world ecosystem by conservation rather than by engineering feats. Is it our duty to help Arizona and California grow or to encourage water-short States to face up to their carrying capacities?

Living things are the only resources that can increase in actual quantity on the globe, for the amounts of replenishable soil, air, sunlight and water—all tied in their year-to-year cycles to geographic position on the Earth's surface—remain relatively constant. Therefore life, and especially human life with its propensity for machine-enhanced gigantism, can outgrow, devour and poison the environment's essential components, of which water is one. We are learning that water must carefully be protected against pollution. Export is a more subtle threat.

The nation-state has this to be said in its favour: it defines one segment of the Ecosphere within which citizens' responsibilities for care and preservation are clearly defined. Look after this part well, and the whole will prosper. The alternative is improvidence at home and eventually disaster for all. Like other nationals, Canadians can trade away their futures by rearranging such environmental mainstays as soils, forests, and especially water, a vital constituent of that part of the world within their political control.

JUNE TRIP TO CHINA, 1983

The best maps of the People's Republic of China in the University of Saskatchewan library predictably appear in the CIA Atlas, showing northeastern China as climatically equivalent to western North America from Alberta to Minnesota.[1]

When in June of 1983 four of us from the University of Saskatchewan visited Changchun, city of "eternal spring," one time puppet state of Manchukuo and now capital of Jilin province, the wide boulevards lined with poplar, pine and, of all things, Manitoba maple, suggested that the CIA had indeed found a match for our regional climate. But at forty-four degrees latitude, the days proved to be mistier, the nights softer, than those I remembered as a youth in Alberta. Also, around the city the well-tended rows of corn, wheat, soybeans, sorghum, sunflowers and melons indicated a Minnesota-type regime: a longer frost-free season, more dependable midsummer rains and altogether a better crop-a-year climate than our prairies afford.

Nevertheless, the soils of the countryside appeared familiar, brown to black in colour, products like ours of steppe grasses and parkland vegetation. But during the Glacial Epoch when the Canadian prairies were under ice, no glaciers touched this terrain, and so the broad plains and gently rolling hills are remarkable to a westerner's eyes in their absence of stones. The soils are residual, deep and cultivatable—new lands for the millions that streamed northward to the frontier, beyond the Great Wall, when Jilin was made a province in 1907 and when the last barriers to immigration by the Han into traditional Manchu territory were removed. Incredible though it may seem, this fertile land was "homesteaded" more recently than our prairies. Indeed, we share a problem of finding how to manage the dry-land farming of grassland soils, without ruining them in the first hundred years of use.

Today, full blown, anticipating the petroleum-poor, over-populated world of the twenty-first century, the horticultural society flourishes. The social focus is squarely on procuring food from the fields. Success is measured by the fact that in little Jilin province, from the foothills of the Chambai mountains on the border of Korea westward over the Manchurian plain, a population larger than that of all Canada is supported on an area less than one third the size of Saskatchewan.

Travelling through the countryside by bus or train, one can easily imagine, through narrowed eyes, that this is our own prairie grainbelt divided into smaller fields by poplar windbreaks but now supporting 360 people to the square mile instead of ten. Each summer morning, a large percentage of the youthful population is off to the croplands on foot or by bicycle, thronging the roads. During the daylight hours, the neat, planted fields that stretch away from the mud-walled villages and tree-lined dirt roads are dotted with men and women toiling singly or in groups, sometimes using oxen or small horses to pull cultivators between the rows but more often weeding around the crop plants by hand and hoe. The typical working dress for both sexes consists of dark trousers and white cotton uppers, though a sprinkling of flashy magenta shirts on the men occasionally turns a smooth green field with its poplar border into the likeness of a golf course.

Machine help on the land in June is not conspicuous; we saw only a few small tractors and self-propelled cultivators. Other mechanization needs apparently take precedence, especially those of transportation. That most appropriate piece of technology, the bicycle—assisting but neither dominating nor poisoning its rider—takes first place, clearly a prized possession and too valuable, at an average cost of two or three months wages, for children's use. Next in numbers on the roads are noisy two-cylinder "walking tractors" and three-wheeled vans, both suitable for carrying medium-sized loads.

The heavy-duty workhorse is a big blue, green or brown four ton "Liberation" truck of International Harvester design circa 1935, hammering along the highways with horn blaring at cyclists and donkey carts, a legacy of USSR assistance that established Changchun as the centre of automotive manufacturing before the Sino-Soviet rift of the late 1950s and early 1960s. Cars buses and motorcycles are relatively rare.

Villages are closely spaced along the roads, invisible at a distance to the untrained eye because they are Earth-camouflaged. A country poor in stone and in wood has only one readily available building material: the soil. Therefore mud, as bricks and plaster or in bulk, is the essential ingredient of peasant homes, animal shelters, and the omnipresent walls that demarcate lanes, kitchen gardens, the outer limits of settlements, and even fields. Walls are ubiquitous in China, perhaps symbolizing millennia of attempts by flatlanders to devise at least minimal protection against invading barbarians.

Houses, typically, are aligned to catch the winter sun, with doors and windows on the southern exposure. The Earthen roofs, usually supported by seven lengthwise poles that protrude above the end walls, are gently curved and serve for the drying and storage of garden-produce out of the reach of wandering goats, pigs and domestic fowl.

As transportation has improved, bringing easier access to fossil fuels, brick kilns have sprouted as a common feature at the edges of towns. The newer communal buildings and homes of well-off peasants are now constructed with mortared-brick walls and pitched, tiled roofs. Nevertheless, the basic material is still the good Earth, quarried and baked into stone. The result of scavenging soil for building purposes at the outskirts of each community is a pitted and devastated landscape to which erosion by animal and human traffic contributes. Overall a lack of colour, architectural flair, of decorative touches, of even the minimum amenity that planted flowers could bring, bespeaks an historic background of drab, no-nonsense, subsistence farming. So it was on our frontier, historians tell us, when John Diefenbaker was a lad.

Although the Manchurian Plain in 1983 may preview agricultural living in the parkland-grassland climates of the world sometime in the next century, few consider it a desirable state today. True, everyone is clothed and minimally housed, people no longer live in fear of starvation as they did a few decades ago, and health care has improved tremendously. But who can be satisfied with the basics when the glitter of a brave new world beckons? "Modernization" is the current catchword, meaning adoption of Western technology as quickly as it can be assimilated, to the end that everyone may enjoy the same range of consumer goods available in Japan, Europe and America.

In the cities, the big department stores are crammed with potential customers avidly examining the radios, stereo sets, cameras, calculators and other electronic gadgetry that will be theirs after a few more years of saving. Out in the country, rural electrification, television and better roads are bringing the values of city dwellers to the peasants, 800 million strong in all the nation.

China is no longer insulated, if the nation ever was, from the aspirations of the rest of humanity. What, then, will be the outcome of one-fifth of the world's people demanding their fair share; a tractor and combine in every field, a washer-dryer and microwave oven in every apartment unit, a Toyota for every family?

The problem is officially recognized, and undoubtedly a go-slow policy is in effect. "In view of China's large population, it is necessary to adopt a more medium-level technology, build more medium-sized and small enterprises, and develop more labour-intensive projects and trades," says the official *China Handbook on Education and Science* (Beijing, 1983).

The spiritual home of Taoism and Mahayana Buddhism has come a long way since the formation of the Chinese Communist Party in 1921. Marxist belief in the primacy of materialistic laws, in the fundamental priority of economic categories, fits communism as poorly as capitalism for dealing with environmental people/Earth problems. Both humanistic "isms" subscribe to domination of nature as a means of advancing the material welfare of mankind. Despite the dogma that historical necessity will bring forth an ethical world from the correct organization of productive forces, exhortations to build a "spiritual civilization" to promote material civilization are still in order. "The role of spiritual civilization is to ensure that modernization will never go against its only aim of bringing [material] happiness to the entire labouring people and not wealth to a few," says the *Beijing Review* (10

January 1983). In a reversal of the gospel of Fabian socialism, the spiritual is to be servant to the material.

Religion is tolerated and is visibly practised in public places. People are free to attend services, burn incense and pray in the temples—albeit often against a background of gaping tourists and flash-popping cameras. Treating religious observances as quaint spectacles may be the one way to assure their demise. Nevertheless, at least one Buddhist of note—Long Lian, president of the Sichuan Buddhist Association and abbess of the Tiekiang (Iron Idol) Temple in Chengdu—blends religion with membership in the Peoples' Political Consultation Conference. "While some Buddhists devote themselves to reaching the Pure Land on the Other Shore, she believes, instead, that one should work to purify one's surroundings and make the real world a paradise," says the *China Daily* (19 June 1983). The gospel of social justice that motivated the clergy pioneers of the Canadian Cooperative Federation burns also in some of the Buddhist sects.

Since Mao's death and the downfall of the "Gang of Four," pragmatism has replaced those rigid centralist policies that now are stigmatized as "left thinking" and attributed to undue influences by the USSR. Lingering sources of embarrassment, reminders of the bad old days, are such things as fading slogans on walls and buildings, the chiming clock of the Beijing Railway Station that plays snatches of "The East is Red," and numerous full-length, outsize, uninspired statues of Mao Zedong in front of or within public buildings. These monumental memorials have a certain aesthetic importance; they prove once and for all that a bust beats a statue any day of the week, for the sculptor able to inject pizazz into baggy trousers and round-toed boots has yet to be born.

The people we talked to freely admitted that the Great Helmsman outlived his heroic years. As long as six years ago, according to the *China Daily* (30 June 1983), Deng Xiaoping declared that the "two whatevers"—whatever Mao Zedong did and whatever he said must always be upheld by one and all—were not in keeping with Marxism-Leninism. The "two whatevers" policy was enunciated jointly in

February of 1977 by the *People's Daily*, the *Red Flag* magazine and the *People's Liberation Army Daily*. But in fact, said Deng, Mao never claimed to be always right. Anticipating the wisdom of Canadian weathermen, he said a man should be satisfied if he could be credited as having been seventy per cent right and thirty per cent wrong in his lifetime.

Examples of "leftist" thinking are combatted daily in the press; for example, ideas that socialism necessarily embraces only one form of ownership, namely public ownership, and that tolerance of individual initiative by the State will surely lead to capitalism. Furthermore, the fight is on against absolute egalitarianism, the new motto being "From each according to his ability, to each according to his work." The old idea that to be poor is revolutionary and to be rich revisionist, we are told, is incorrect.

Rapid development of the productive forces in villages under the "responsibility system" proves the evils of egalitarianism, say the *China Daily* (16 June 1983), for payment linked to production has brought economic development. In a country whose peasantry has long lived in poverty, whatever promotes the growth of productive forces is best. The point is illustrated by the story of Song Zeming, a peasant who last year made a large profit by growing sweet grapes under the rural production responsibility system. Worried about what his cadres might think, he kept secret his "honest-gains-through-labour." When, however, General Secretary Hu Yaobang reported to the twelfth Party Congress that the responsibility system would continue for a long time, Song Zeming felt relieved and made public his income of seventeen thousand yuan (CAD $10,635). "The change in Song Zeming's way of thinking is typical in China's rural areas today," says the *Beijing Review* approvingly.

The fine line between enterprise in the service of the socialist state as opposed to enterprise in the interests of the individual or group is difficult for some to draw. Consider the unhappy case of Yu Youqing, whose selling of sub grade tomatoes 30 per cent higher than the official rate triggered the Beijing authority's decision to reorganize his greengrocer shop in early June. As reported, Yu had thought that over-charging the customers would bring more income to the shop, while the buyers were willing to pay a little more to get the tomatoes anyway. After two weeks of re-education he was back in business, warmly welcomed by the neighbourhood residents who will, however, "be watching to see if the reorganization will prove truly successful." (*China Daily*, 28 June 1983).

More serious are "the unscrupulous cheats profiteering under the guise of implementing China's economic reforms," according to *Renmin Ribao* (27 June 1983). While material interest is the motivating force for all social activity, said the paper, under socialism this embraces not only the personal interests of the worker but also the interests of productive enterprises, and particularly the general interests of workers and the multi-national citizenry represented by the State. Using a simile that bird lovers would question, and particularly Canadian conservative bird lovers, the *Beijing Review* likens the relationship between individual enterprise and the common weal to a bird in a cage: a bird should be allowed to fly, it says, but only within the framework of a cage; otherwise it will fly away. One suspects that on occasion the bird has flown the coop, to Hong Kong and on to Vancouver.

In China, as in British Columbia, the tenure system for government employees, including university faculty, is under attack. Job guarantee for a lifetime regardless of performance, a system called "The Iron Rice Bowl" (the bowl never breaks!), does not fit well with the doctrine of distributing rewards according to useful work performed. Regrettably, many employees of the State—including academics, who are protected by the Iron Rice Bowl—simply goof off; they consume but they do not produce. "Why not abolish the system?" asks the *China Daily* (16 June 1983), adding that such an action would only affect 83 million people, a mere 19 per cent of the total labour force. The famous quip of C.D. Howe, "What's another million?," springs to mind.

Indeed everything turns on numbers in China. Mao Zedong

encouraged large families, and one of his legacies since the liberation in 1949 is a doubled population with 55 per cent younger than twenty-five years of age. Everyone is aware of the population problem, of the more than one billion to which 15 to 20 million new consumers are added yearly. Drastic measures have been adopted, such as discouraging families by social and economic pressures from having more than one child (except in the borderland autonomous provinces where, presumably, large families mean more defenders and greater future security). Still, even with family planning and easy access to birth control including abortion, it will be many years before population growth is curbed and numbers stabilize.

As the rules and regulations in cities attest, large population-to-land ratios necessarily entail social restrictions on the individual. Nearly one hundred years ago William Graham Sumner, a Yale sociologist, argued that democratic institutions are only in part ideational and traceable to a particular belief system. The actual existence of democracy, he said, requires favourable ecological circumstances, especially a low people-to-resources ratio. According to this hypothesis, only rich countries—or what often amounts to the same thing, thinly populated countries—can afford to place a high value on the individual, on freedom and the philosophy of democratic liberalism. One ought not to be surprised that China—densely populated, historically inturned and bureaucratic—has adopted a political system with strong centralized social controls. The dangers of concentrated power were manifest when ideas at the top changed in 1957, 1959, and most recently in the period 1966–76.

In the two earlier years, intellectuals (loosely defined as any with above-average education) were accused of being "rightists" and were singled out for attack, while during the latest episode, the Cultural Revolution, they were persecuted as "bourgeois revisionists." The "leftist" is today's bête noire.

Population pressure also fosters a utilitarian outlook, and this has come to the fore under the leadership of Deng Xiaoping. Such slogans as "Seek truth in facts" and "Practice is the only criterion for truth" direct attention away from doctrinaire theorizing. In many endeavours, practically seems to be close to the surface. Thus, for example, no bones are made about the aims of education: to inculcate love of country, love of socialism, love of the proletariat, as well as to speed modernization.

Forthright statements, right or wrong, about why nations place a high value on literacy and learning are praiseworthy. Parallel aims in our educational system are muted or disguised, doubtless because we have never gone through a revolutionary wrench requiring an appraisal of directions. Smug and insouciant, we seldom ask what Canadian schools and universities are doing to us and for us.

The Cultural Revolution was a remarkable convulsion, an attempt at ideological purification by turning power back to the proletariat. Workers, peasants and soldiers were encouraged to administer and reform secondary education. More than 100 institutions of higher learning were abolished or broken up. Faculty were dispersed, many being "sent down to the country" to be re-educated by manual work. The universities are still trying to get back their buildings and facilities. They are attempting to recover the lost decade by expanding contacts with foreign institutions.

Our mission, as a four-man team, was to arrange an exchange between Northeastern Normal University, its Biology department and the University of Saskatchewan, with a particular view to fostering ecological research on degenerating grasslands. From now on, for many years, scholars and post-graduate students will be trading time between Changchun and Saskatoon. We hope to learn from their experiences, and they in turn will search the prairie-province scene for transplantable modernization.

Hospitality abounds for Canadians visiting China. Norman Bethune continues to be our greatest ambassador of goodwill. But even without his commendable exploits, mutual exposure would surely make us friends. For the Chinese temperament is open and warm, appreciative of humour and, we found, of Western music. A favourite song, from Changchun to Shanghai, is "Red River Valley."

CHINA REVISITED, JUNE 1985

All through the first week of June the snows fell softly in the city of Changchun. The main thoroughfare, Stalin Avenue, was covered with billowy drifts through which the citizenry poured by truck, bus, bicycle and on foot, stirring and scattering the white stuff. Doll faces of garishly clothed tots riding up front between the arms of cycling parents were protected with fine nylon windscreens tied over the hat and under the chin, and the more fastidious women encased their heads in a similar fashion. Sight of the white ground cover under the greenleaf trees fleetingly brought back memories of the untimely onset of winter on the prairies in mid-October, 1984, injecting a psychological chill into the warm summer air. For this was only fake snow—poplar fluff—a legacy of post-Liberation landscaping enthusiasts who lined both sides of the long avenue, far as the eye can see, with female seed-producing trees.

Such unexpected reproductive bounty from a boulevard community assumed to be sexless is not unknown in Canadian towns and cities too, but in populous Jilin province it assumes a certain portentousness, a warning sign that fecundity will out. For China is trying to downplay the importance of procreation and of the corollary that sex is consequential. The nation's present population is estimated to be 1.03 billion with a preponderance of young people, and the optimistic goal is to hold the demographic line at 1.2 billion by the year 2000.

Billboards showing handsome couples gazing into the star-lit distance in a misty ambience associated, in the West, with the sale of body lotions or condoms are here vehicles for such chaste messages as "Let's Postpone Marriage for Awhile." Those already married are encouraged by the roadside slogan, "It is Glorious to Have Only One Child."

Close control of public advertising seems to be helpful. No billboards or magazines trumpet the North American refrain that sex is as central as heating. The typical TV entertainment of an evening, based strictly on interpretation of sights and sounds, recounts the saga of a family or group of patriots escaping from feudal oppression, the men shouting, the women sobbing. Background music for such drama is heavy on violins and I'll-take-you-home-again-Kathleen themes. No raucous saxophones tempt Katie to take it off.

A newspaper poll conducted in one of the China's teeming provinces asked married couples the question: "Why did you marry?" Responses were compiled in five categories of which, the report noted, the government condones the first three: for love, for balanced considerations (meaning with sober and commendable attention to economic factors) and for traditional reasons (meaning arranged by family). Viewed with disfavour by the government were the fourth and fifth: for convenience (to obtain some calculated advantage of living conditions or employment) and for the satisfaction of sexual needs. Fortunately only 6.5 per cent of respondents identified with the last, leaving a healthy 93.5 per cent who presumably couldn't care less.

A more revealing category, prudently omitted, would be one labelled "in order to raise a family; to have children." The Chinese love big families, a native son told me, and the one-child-per-couple rule is very hard. Out in the country where 80 per cent of the population still supports itself, the rule has been especially unpopular and public pressure has recently led to its relaxation. Rural families with "special difficulties" are now allowed to have two children and members of ethnic minorities can have three or more.

"Special difficulties" concern the problems of looking after the aged in an agrarian society that as yet lacks the resources to provide pensions, social services and housing for the retired. Children, and especially male children, are still the only old-age insurance in sight. By the end of this century when nearly 11 per cent of the population will be over sixty years old, will there be enough offspring to look after 130 million oldsters? How will two young married people, each an only child, support four aging parents, their own child, and perhaps a couple of long-lived grandparents?

Nevertheless as late as 1985 the *Beijing Review* was enthusing that country people have seen the light and have recognized that the first national priority—raising the productivity of farms—does not go well with raising children. Zhao Shujun, a rural woman in Tianjin, would have a difficult time handling her money-making pigs and chickens if she were to have more than one baby, observes the reporter. Yu Zhencheng and his wife banked 6,000 yuan (CAD $3,000) last year, which they could not have done with two children to support. At this moment when "Time Means Wealth," a wise peasant will not lose the good opportunity to get rich rather than supporting several children. And as the clincher, "Single-child families have 10 per cent more TV sets, 28 per cent more electric fans, and 60 per cent more washing machines than the average local level."

At the moment, an optimistic mood prevails in China. After the lapse of two years, a return visit by several of us from the University of Saskatchewan to Northeast Normal University in Changchun revealed many improvements: new buildings going up on campus and an impressive new gate at the entrance, signs of cleanup and painting of class rooms and other facilities that suffered neglect during the Cultural Revolution, and increased academic activity in response to national needs, including the training of more graduate students.

Our associates at the University were friendly and helpful, solicitous as always of their visitors' health and comfort. The atmosphere was good, and the undercurrent of tension that worries Westerners in some of the other centrally planned nations was not apparent. At least in the circles where we moved, there was no hint of surveillance, and, considering the trials and stresses that have beset "intellectuals," the university people appeared to be as relaxed and good-humoured as academics anywhere.

In the city itself, as elsewhere in China, the economy is booming (some say to the point of overheating) and change is in the air. By opening its doors to the world, China is confronted by the problem of separating what is useful and nourishing from that which may prove indigestible—not an easy task when the Western fare offered is so rich and varied. The country is a swarm with experts and entrepreneurs, and all are welcomed warmly and listened to attentively as if in their baggage they carry the academic and technical answers for a nation eager to overcome past isolation, to catch up to the over-developed West, to be again the Middle Kingdom at the centre of things.

Attesting to the importance placed on visitors to the campus is a University Department of Foreign Affairs, kept busy organizing a daily round of welcoming and farewell banquets as foreign professors come and go. Among the experts we encountered casually both on and off campus were electronics specialists from Japan, English teachers from the UK and the USA, an Australian irrigation agriculturist, a Dane skilled in furniture manufacturing, a German Egyptologist, American sociologists and historians and a theoretical physicist who endeared himself to me by opining that China has far greater need for ecologists than for people with his type of learning.

Modernization is the official goal, and all visible signs suggest that to mean "Westernization." Since my last visit, clothing, transportation, music and building construction have undergone dramatic conversions.

The drab unisex blue or green pants-and-tunics costumes are giving way on the streets to clothes more varied in style and colour. Now on a Sunday in the park, the well-dressed young woman trips along on moderately high heels, wearing a bright blouse with either designer jeans (which will be skin tight next year but not quite yet) or a knee-length skirt (occasionally even a daring Suzy Wong slit skirt). Her boyfriend sports a shirt and necktie with suit or slacks and jacket. His leather shoes or sandals have platform heels, and on the ankles of his nylon socks a lacy pattern is embroidered. If clothes are any indication, sex is alive and well despite official polls.

As to transportation, roads that two years ago served mainly as arteries for bicycles, with horse, donkey and bullock carts plus the occasional snorting truck and tractor, are now thronged with newly imported smooth-running buses, vans, trucks and taxis. Overnight the sturdy but sluggish vehicles produced by communist countries have been rendered obsolete by an influx of Japanese-engineered automobiles. Along with them a whole new generation of young kamikaze drivers has appeared, trained to pilot Toyotas, Nissans, Mazdas and Mitsubishis, as adept at raising and lowering windows at the touch of a button as at flicking on the syrupy pop-music tapes even as the key is turned in the ignition.

Nostalgia for the Old China of 1983 rose from my heart and gathered chokingly in my throat on a misty July morning as I was hurled down Beijing's new Third Ring Road at 120 kilometres an hour in an air-conditioned red Toyota, the balanced stereo speakers piping a jazzy Japanese version of the Toreador chorus. Later that day in Hong Kong, I was treated to a similar low-altitude flight by a Nissan cabby between the airport and downtown Kowloon. Like his opposite number in Beijing, he too affected dark sun glasses and spun the wheel with cutaway racing gloves. The year 1997 promises extraordinary carnage on the Kowloon peninsula when the left-hand drivers of Hong Kong are merged with the right-hand drivers of the China mainland.

But by far the most remarkable change in the People's Republic has to do with construction. Everywhere, in country and in city, the cultural landscape is being refurbished in a colossal Earth-moving and building spree. Perhaps of all people, the Chinese are most adept at manufacturing what they need from the planet's skin. Inured to a wood-poor milieu that many millennia of intensive agriculture has created, they learned long ago to exploit the very Earth itself in order to build great hills and burial mounds (tumuli), walls and fortifications, homes and public buildings.

Lying everywhere today, by streetside, roadside, or in transit on innumerable trucks and carts, are piles and heaps of rubble, quarried stone, lime, sand and bricks. And everywhere lie the pits, quarries and ponds from which the raw materials have been or are being extracted. Draglines, building cranes and brick factories are the badges of the new China.

At the outskirts of the ancient city of Xi'an on the Loess plateau at Banpo, a Neolithic village site discovered in 1953 has been excavated. Clear evidence has been found of square buildings and halls built of clay on frameworks of poles and wattles, a direct link from six or seven thousand years ago with the rectangular plan and construction materials of contemporary peasant homes. But these latter will soon be museum pieces too, like those at Banpo. Kilns and their smokestacks dot the landscape, and the mark of a successful peasant is a fire-brick house with a red tile roof. Urban renewal also means the demise of the mud-brick huts and shanties. Rising in their place to accommodate a denser population are dreary concrete shoe-box apartment blocks, row on row, demonstrating once again that cheapest is ugliest.

As a footnote to Banpo, its exhibits illustrate the theoretical slant of historical interpretations. The meaning of the past as promulgated by group or nation can reveal at least as much of current ideology as of the events it is supposed to mirror. This Neolithic site tells the Marxian tale of social evolution whose motive forces are the changing means of production. Piecing together the notes on the various exhibits, the following story emerges:—Early society (stone age) was meager in means of production; it was matrifocal, classless, practicing a primitive communism, with religious ideas that grew out of ignorance about life, death and the phenomena of nature. (Science renders religions obsolete by demolishing their foundations of ignorance.) Then women invented agriculture which contributed to their downfall. The increased means of production broke the relationship of equality between the sexes and allowed men to take control of property. Unequal wealth created classes and class conflicts, from whose contradictions the institution of the State emerged.

At this point, the Banpo historians come to a full stop, with no prognosis of when the State will wither away. Details of the story are filled in by the sometimes forced interpretations of particular artifacts. Fine art work, for example, that might be supposed to reflect the inspiration of some lonely genius, displays instead "the outstanding creative power of the labouring people." Again, the idea that progress necessarily means mastering nature—an essential component of Marxian belief—is gratuitously introduced or grafted on otherwise reasonable statements. Thus the use of bow, arrow and spear "lengthened man's ability to conquer nature," while a fish-mermaid design on a clay pot "reflects their desire to catch fish, and their strong determination to conquer nature." Doubtless our historical narratives seem just as manipulative to the Chinese.

Save for the cheerfulness and bustle of their inhabitants, the cities of China are shabby and depressing. With the disappearances of the emperors, both aristocratic tastes and the wealth to make something of them also disappeared. Ever since the last of the Manchus was forced to abdicate in 1912, constant wars and revolutions, along with grinding poverty, have suppressed artistic architectural expression. Apart from the treasures of the distant past—the mausoleums, pyramids, palaces, statuary and formal gardens of the Tang, Ming and Qing (Manchu) dynasties, and the temples and pagodas of Taoism and Buddhism—the urban centres show little in the way of flair and style. This is especially evident in the new industrial cities of the Manchurian Plain in northeast China where the provincial capital cities of Harbin, Changchun and Shenyang exhibit few examples of indigenous architecture. In a sea of factories, tenements and smog, the best they have to offer are the monumental but tasteless public buildings erected by Japanese invaders or Russian helpers.

The lack of urban attractiveness may stem in part from the determined rejection of a historic past whose every aspect, until Liberation in 1949, is associated with feudal tyranny and unworldly religious faiths. The Cultural Revolution, 1966–76, is the latest reminder that the past is best forgotten quickly, and that little except the technologically new merits attention. The history that commands intense interest among the new generation chiefly concerns the rise of the Party and the establishment of the People's Republic.

A university student took me to the site of the ancient imperial resorts built around hotsprings at the foot of Black Horse Mountain in Shaanxi Province. Records show that the first palace was constructed there during the Western Zhou Dynasty, some 900 years BC. The first emperor of Q in (or Ch'in, from which the name China was derived), he of the famous terra cotta warriors, also had a residence there more than 200 years BC, but the pleasure domes and palaces on which the present pavilions and gardens are based were designed by a Tang Dynasty architect in 644 AD.

While my interest was mainly in the halls and the baths of the Glorious Purity Palace where the Tang emperor cavorted with his favourite concubine Yang Guifei, my guide's attention was rivetted on the bedroom just behind the Imperial Concubine' s Bath. From it, on December 12, 1936, Chiang Kaishek fled in panic up the mountainside when his troops mutinied in protest against orders to fight communist compatriots rather than Japanese invaders. Later that same day, the generalissimo was found ignominiously hiding in a cleft in the rock, minus one shoe and his false teeth. This, the "Xi'an Affair," draws the proletarian crowds to the hot springs.

Rejection of things feudal and ancient extends also to the philosophies and religions that Caucasians often associate sympathetically with China: the endemic Taoism and Confucianism, and the Buddhism transplanted from India about 100 BC when a commercial route to the West was opened along the Silk Road. But doctrinaire communists find objectionable precisely those aspects of the Eastern philosophies that appeal to our semi-ecological, semi-postindustrial society: a simplified lifestyle, detachment from material possessions and a let-it-be attitude to the natural world. Like the muscular Christianity that is its inspiration, communism is set on changing the world, not accommodating to it.

As with the Buddhists who came later, old Lao Tsu taught that the Way (the Tao) lies in seeking freedom from desire and from sensory experience of the "ten thousand things" which is illusion:

> The Tao that can be told is not the eternal Tao.
> The name that can be named is not the eternal name.
> The nameless is the beginning of heaven and Earth.

The named is the mother of ten thousand things.
Ever desireless, one can see the mystery.
Ever desiring, one can see the manifestations.[1]

For awhile, during the Sung Dynasty, the Confucians opposed the meditative techniques of self-improvement advocated by Taoism and Buddhism, but in time they too turned inward and finally became ultra-conservative, endorsing the belief that the status quo represents the highest order. And so, according to critics in the People's Republic, all three faiths are debilitating, selfish in their attention to personal rather than social improvement, opiates that foster apathy and indifference to modernization and change.

Not surprisingly the faith invoked as a virile substitute for the decadent old ones is Humanism, defined as active intentions to increase material well-being, dissatisfaction with the present, openness to the world and hope for change, rejection of the Confucian doctrine of the mean, emancipation from the taint of asceticism, development of the personality by desiring and enjoying.

If this sounds like a neo-Victorian materialistic philosophy suited to a rapidly industrializing country, why so it is. The feeling soon grows in a visitor that the history of the 19th century West is repeating itself in China. The factories and belching chimneys, the smoky streets and dusty tenements, the building debris, the polluted water, the peasants arriving from the country with their bedrolls, these all bespeak the times of which Dickens wrote. China's intention to "catch up" is telescoping and concentrating both the benefits and evils of industrialization that in European countries were spread over more than one hundred years.

Despite ideological differences as to who should control capital in society, socialism and capitalism share the same vision of a future in which all humans will be wealthy and omnipotent, and from those two supposed virtues also healthy and wise. The humanistic faith of both is in science, technology and "the innovative genius of man." This philosophy places high value on quantification, efficiency and mechanistic control. Both systems are frightening because in their pursuit of power, euphemistically called "development," they smother or displace those person-to-person values and person-to-nature ethics of the gentler philosophies now outmoded.

Among the famous Ming Tombs near Beijing, a golf course is under construction. Citizens of Xi'an are pondering whether building a Disneyland by the Qin tumulus may not be the best way to attract more tourists. Middens of sunflower husks, hard-boiled egg shells and pop bottles are accumulating along the sides of the Great Wall. Opinions differ as to whether two or five of the beautiful Manchurian tigers still survive to roam the mountains of the northeast.

One of the five holy mountains of Old China is Changbaishan in eastern Jilin Province, on the border of North Korea. It is a high volcanic peak within whose cauldron lies a shimmering lake three hundred metres deep, blue-green like Kalamalka Lake at Vernon. Here the Seven Radiant Daughters of the Sky Emperor have been known to bathe. The shores are littered with trash, left by visiting pilgrims.

"When we had political differences with the Koreans some years ago they were not friendly to us. They objected to the littering on our side of Heaven Lake," the guide told us. "At that time we kept the shore clean. Now that we are on good terms they don't complain, and so the lake is suffering."

"Why did the Koreans object to the pollution of the lake on the China side? Is the lake sacred to them? Are they religious people?"

"No," said the guide with a wink. "They're like us; they don't believe in anything."

Not quite true. For set high on new city buildings, amid the bamboo scaffolding, the maxims "Time Means Wealth" and "Efficiency is Life" have replaced "Heed the Sayings of Chairman Mao." Viewing these new articles of faith, long the mainstay of capitalist culture, one wonders what the authorities have in mind when they exhort their people to avoid cultural contamination, to root out unhealthy tendencies and to maintain the spiritual purity of socialism.

On the train between Changchun and Beijing, I met a Chinese

man-of-the-world, a blasé entrepreneur recently transplanted from the mainland to Hong Kong, planning to build in Beijing the biggest tourist hotel in the country. He entertained us with some of the latest gadgetry: a pocket radio with separate diminutive stereo speakers, a magic cigarette lighter. I asked him about the future of the nation. "What will modernization do to China, say by the end of the century?"

But even he, bellwether of the future, betrayed uncertainty and doubt as he paused a long moment before replying with another question. "What is possible in this country," he mused, "when there are so many people?"

ENVOY

MODESTY IN THE HOME PLACE

The common theme of these essays is that you and me, we people, have evolved, grown in numbers and intelligence, developed values and cultures, all within a nurturing global Ecosphere that is neither well known nor deeply appreciated. The ecological truth that we are parts of a surrounding mysterious whole has not been an axiom of our thinking and feeling. Not realizing that we are Earthlings with the responsibilities and rights that living well in the Home Place entail, we have tried to anchor our natural sense of belonging in fables of special creation, arrogating God's image as the mirror of ourselves.

Blind to our part-to-whole relationship and setting ourselves above and opposed to the natural world in which we are embedded like cells in a larger living body, we have brought down on our heads environmental nemesis that is the penalty of ignorance and vanity.

All is not lost. Time has not yet run out and we can do better. The important institutional changes will come quickly when enough of us realize in a fundamental way *who on Earth we are*. Then the reverence for life that Albert Schweitzer advocated will be extended to embrace the entire planet as we begin to understand our just place in it. For life, vitality and creativity are properties of the whole Earth, the Ecosphere, and not just of selected organic components on which mistakenly we have pinned the "alive" label.

That attitudes must change is a truism. More to the point, what must we give up and what can we retain from the old traditions and their ethical systems? How shall we come to terms with the ecological insight that people are not wholes but inseparable parts of the natural world, and that when parts attempt to dominate the Earth-whole they cause pathological problems? Answers require an examination of cultural beliefs about the Nature that surrounds us and a confronting of their errors. Consideration of what we know about our minds and the world is the place to begin.

Attempts to make the mind intelligible can tie the mind in knots. Nevertheless those who have pondered on it conclude that our minds are active and creative—otherwise information coming in through the senses would be disorganized and chaotic.[1] The first creative accomplishment of mind is the world itself, the world it constructs from the buzzing, blooming confusion of stimuli with which from birth we are bombarded. Some unknown mind-quality affiliated with ourselves, but not necessarily so confined, builds mental models of a universe into which we enter as our common-sense world, finding it furnished (though we have had a hand in the interior decorations) with constructs such as "mind" and "body," "spirit" and "matter." Rather than accepting these constructs as artistic accomplishments, we objectify them as realities that exist apart from our perceptions and set out with the confidence of science to explain what makes them tick! We are naive realists chasing will-o'-the-wisps that may or may not prove to be the light.

This is not to discredit all the perceptions that our culture shares—for we must hold together and give allegiance to some common reality—but to suggest that other realities are entirely possible. Whatever the authenticity of Carlos Castaneda's experiences, each of his books conveys the truth that different cultures are different realities.[2] Thinking changes, cultures change, we change and our thinking changes—an evolving spiral through time. The challenge today within our imagined world is to imagine a future that leads away from the problems created by narrow thinking, a Way that will make our culture more becoming and more fitting—which is one definition of a religious quest for the Tao.

To fuse the old idea of Nature with the new one of ecosystem is a constructive step. The Nature with which we are in direct contact is the Ecosphere, the entire global ecosystem: air above, water and land below, ourselves clustered with other organisms at the gas-liquid-solid phase boundaries. In the interests of contemplation, study and understanding, we can divide this largest ecosystem into smaller sectoral ecosystems in descending order of size and inclusiveness: the

North American continent, the Great Plains grass lands region, the Cypress Hills, a field, a lake or a town. All are vivacious Earth-space volumes encapsulating organisms, and all are interrelated parts of the whole. Within them we live, move and have our being. From life within one ecosystem we know something of the others, and after a single generation in a new place, anywhere in the supportive matrix, we soak up a sense of belonging and name it our Home Place.

The concept of Nature-as-Ecosphere—comprising complex, diverse, vital, evolving volumetric ecosystems at the Earth's surface—is exceedingly important because it places in perspective all the component parts—including people and their cultural institutions. Traditionally we have thought of humans singly and in social groups as the two most important realities and on this confined base have erected competing ideologies of liberalism and socialism, anarchism and nationalism, free enterprise and the welfare State, personal salvation and the social gospel, psychology and sociology. A central problem of politics has been the reconciliation of person and community, of the individual will and the general will. Political parties define themselves by the emphasis given to the individual versus the collectivity, We should be exploring with equal zeal the dialectic that ties together people—individually and collectively—to the Ecosphere, ourselves parts of it, and it the extension of ourselves. The strong and melodious voice of a third member—the Ecosphere—can harmonize the dissonant pipings of the individual and the collectivity. A trio is the potential cure for a weak duet.

Poorly formulated and misleading ideas of ecosystems and ecology as pertaining only to biological communities have recently appeared both as targets and artillery in the people/society debate. Those eager to defend aggressive individualism smell danger in an ecology assumed to emphasize the importance of "biotic groups," for if the latter are "morally significant," they detract from the moral significance of the individual. When ecology is confused with sociology and equated with "biocentrism," it becomes the bogey-man of liberalism and of those whose first concern is maintaining the autonomy of the individual.[3] On the other hand, those attracted to communitarian principles eagerly recruit ecology as an ally, defining it as the organizing principle of social development, as "social ecology" concerned with ecocommunities.[4] The inclusive ecosystem concept is misread, for Nature is neither "biotic groups" nor "ecocommunities" but total Earth-space in which organisms are just one ingredient among others of equal importance.

Such examples illustrate both the one-species focus natural for our self-conscious species and the fears of what might happen to our cherished egos should that focus shift. What will become of the human race if it takes its eyes off itself, if it ceases to think only of *Homo sapiens?* Cut away our traditional worldview and what of worth will remain in the libraries of our institutions and of our minds? What cultural goals for society in politics and economics, what religious and scientific exercises, could possibly fill the vacuum if the old anthropocentric ones are carted away?

We can take heart that our race survived an ego-crushing blow four hundred years ago when, in the Copernican revolution, an Earth-centred universe was regretfully relinquished. Once again events are forcing us to let go of cherished and cramping misconceptions, and loss of a people-centred Earth will be just as tough to take. With free fall from the second Copernican revolution on the way, any kind of parachute is welcome. Unfortunately, contemporary writings offer little in the way of rescue: no quick life-savers and no new goals and directions to allay fears of losing that on which we have hitherto staked our lives.

PLAYING CAPTAIN OF THE SHIP

Today's lack of direction partly results from our being overwhelmed, in the space of a few decades, with a mess of complex environmental/social problems. The first order of the day is fighting brush fires—cleaning up pollution, organizing food banks. After that we intend to get around to asking what started the fires. A distant third is the question about what we will do differently if we manage to put out all the brush fires, yet an answer to this question may be the burning solution.

Our minds tend to work backwards. We sail along until we encounter an obstacle, a problem, and then ask how did we happen to run into this? Search for causes turns the mind toward the past, although the problem might have been avoided had we set a different course in the first place. Thorough diagnoses and historical analyses of the human condition, exploring problems and baring their roots, are necessary and useful. They ought, however, to be balanced with imaginative visions that give form to the future. To prevent worse woes, looking ahead is more important than analyzing the past.

Not that we are without general humanitarian directions and answers to perplexing questions. Yet most are cast in the old mold: we will make a more prosperous and just society, a kinder gentler society, a more humane society where animals are accorded better treatment, an environmentally aware society in which pollution is a crime. Such thoughts express admirable moral goals though they are a bit nostalgic, backward-looking, stereotyped and static. Somehow they do not connect with visions of an exciting new world with beautiful people in it.

Well then, why not the exhilaration of teaming ourselves with process, with change, to engineer a New World closer to the heart's desire? After all, we *are* different from the rest of creation, we *are* special as well as part of nature. Whatever we choose to do is natural. So why not pursue our science and technology enthusiastically. We live with change: the universe expands, the dynamic world shifts, climate changes, and in the midst of it all people have become a geological force greater than Earthquakes and volcanoes. We have the power to remake the face of the Earth and are now guiding organic evolution through biotechnology. Why not be honest about ourselves, accept our manifest destiny and wholeheartedly embrace our roles as both pilot and captain of the good ship Evolution and as co-creators of the universe?

The arguments are seductive. What a challenge to govern evolution creatively! We ought not to be left out of the creative process, nor need we be if we simply recognize our power at the centre of it all. Nature needs our help, and she can be improved for our own betterment. Let's get at it with more money for research, bigger laboratories. Engineers and technicians to the fore!

But wait a minute. To take over the reins of evolution must mean that we know where we want to go. If we steer evolution "for our own betterment" we should first know what is good for us, or at least which direction is best. And is not this precisely our problem? We have no worthy goals, except those that a selfish and exploitive and careless species has wrongly thought were in its own interests. The very environmental/social problems that beset us are the result of a failed world view focused on the human race. Reinforcing it by taking charge, "for our own betterment," will do us in more quickly, not make things better.

AN ECOLOGICAL GOAL

Who are we on Earth? The paragon of animals, conscious, imaginative, creative. The animal that can most destroy, the animal that can most change. We have the power to make choices and act on those that are preferred. Could we but understand how we connect and are joined to the Ecosphere, might we not find a standard beyond ourselves against which to judge between better and worse courses of action? In the past, religious teachings have tried to provide such standards, helping people to transcend their own immediate desires, their self-centred urges. The insights of ecology offer an additional universal standard, at least for life on this Earth.

The key idea is this: Nature, the Ecosphere, is supra-organismic, not an organism but something even more complex, more interesting and more beautiful: a higher level of organization than things like ourselves.[5] We flourish and thrive within this supra-organismic Being, one of her components that busily breathes oxygen and burns carbon. We are inextricably a part of the Earth-Nature-Ecosphere without which we are nothing.

Within Nature's body we are like the free-living cells within ourselves. Had they our same level of intelligence they too would ask: What is our goal? The obvious answer would occur to them: your purpose is to grow, reproduce and live happily to a ripe old age. But from the broader wisdom of our vantage point, knowing ourselves to be the embodiment of the cells, we perceive that their health depends on our body's health. Should they turn pathological either in numbers, in demands or in the production of toxins, then the body will suffer and so will the cells. The foremost goal of the cells, then, and the secret of their inborn rationality or instinct, is to assure the health of the body. The parts serve the whole. Is this not the function, niche and role also of organs: the heart, kidneys, brain?

Just so, the human species and all like animals and plants represent a lower level of organization than the Ecosphere that produced and maintains them, suggesting that their primary role, purpose or niche is to maintain the health of the Nature in which they find themselves, to attend to that which is more important than they—judged by precedence in time, complexity and creativity. Health, haleness, wholeness and holy are all related words, reminding us that to keep the world healthy is to keep it whole and holy. Further, beauty is an attribute of health and a frequent expression of wholeness—the relating and integrating of diversity. We should expect a healthy world to be also a beautiful and diverse world.

The new goal, *Guard and Maintain the Health and Beauty of the World,* encourages humility, a virtue in short supply among us. Its pursuit will keep us healthy, the first attribute of beauty. It sets a premium on quality as opposed to the goals of quantity that are killing the Earth and destroying our civility. A safe and constructive purpose, it redefines progress and provides standards against which we can judge how well we are doing with our population numbers, the quality of our industry, the directions of our science and technology, the appropriateness of our education and recreation. It extends Schumacher's goals for good land use to the entire globe. Guard the health, beauty and permanency of the land, he said, and productivity—the quantitative aspect—will look after itself.[6]

Religio, the root of religion, means to bind together, to experience wholeness. What is bound together in the religious experience of oneness may be persons in small social groups, in large communities, ecosystems, nature, the universe, the ineffable Tao. The achievement of each enlarged unity enlightens the lesser ones. Mystics who have attained the ultimate Unity have not been led by that experience to renounce their community ties but, on the contrary, have usually strengthened and enlarged them. The saints have not come down from the mountain hating humanity but loving people more. Nevertheless it is commonly supposed that experiencing the sacred in Nature, venerating the greater-than-human Ecosphere, will somehow be anti-humanitarian and anti-religious; to love the planet more will mean loving people less and perhaps being godless?

Some have argued that we must be right with ourselves and our society before we can be right with Nature. Men learned to be cruel to Nature, so the argument goes, by first being cruel to women, children and animals, and only when the human love-bond is strengthened will they learn to love Nature; as long as psychological problems bedevil the individual and injustice rules society, so long will we bedevil and unjustly treat the world. This is exactly the wrong way around. In theological terms, the proposition is foolish for, in effect, its logic suggests that we must love ourselves and our communities before we can love anything higher, before we can love God. Charity is no respecter of boundaries and neither love of the Tao nor the love of Nature requires apprenticeship in a Service Club.

Most of the disputation comes down to this proposition, that we cannot bear to give up our individual and group egoism. We have mesmerized ourselves with talk of our preeminence, our specialness, our uniqueness in the animal world. We have made ourselves in God's image. We have declared our "innovative genius" and expressed pride in being as we are, asserting that we will be "less than fully human" if we curb our assertiveness. To whatever degree this is true, it is dangerously unbalanced by humility. The conceit and arrogance of our species explains why we are such dangers to Nature and to ourselves.

Yet we have our moments of enlightenment. Collectively and recurrently we show our truer colours, banding together to accomplish worthy goals, freely giving in the interests of a perceived higher good, responding to prophetic visions in times of crisis. One such time is now upon us, and its demands are not excessive. What the Home Place needs from us is more modest furnishings, less extravagance, more tender loving care. The appropriate vision is Ecumenical Ecology, and both words remind us that it is time to come home.

BIBLIOGRAPHY

Each of us comes to her/his philosophical position through a lifetime of picking up ideas, mostly at random, constructing from these bits and pieces a more or less consistent map of reality. Basic to the process is a set of key concepts, a general theory or paradigm of how things are, around which our experiences are progressively crystallized. When our "crystallization centres" are different—and none of us shares with others exactly the same world view—then what we distil from a particular author may, in varying degrees, make sense or not.

The following writers make sense to me as I hope they will to you. They contribute to a perspective that encourages us to take our eyes off ourselves in order to appreciate a wider world. All are critical of the limitations of the homocentric ethic. Although they may not explicitly conceptualize the Ecosphere as the immediate trans-human reality of supreme value, they point in that direction.

The best place to begin is with the writings of the poetic naturalist, John A. Livingston, whose many books and articles play variations on the theme of escape from the false notion that "man is the measure of all things." Our madness, he says, is the hallucination of human power over everything non-human. Neil Evernden's related ideas expose the source of our alienation from the natural world. Alan Watts uncovers the foolishness of the convention that each of us is separate from what surrounds us; in reality, the World is Your Body.

Fritjof Capra, physicist-turned-philosopher, shows how the 400-year idea of a mechanical universe separate from humanity is yielding today to a more organic vision of reality that revives the ancient motif of a feminine nature. Caroline Merchant develops the same theme, pointing to the intensely masculine agenda of science and technology as mastery and control over nature. David Ehrenfeld exposes species-centred humanism, the world's most popular religion, unrecognized as such because its rituals are our daily routines of business and community service. Joseph Meeker points out that the heroes of tragedy are the wrong role models for a world suffering from human self-importance. The characters of comedy teach us compliance, tolerance and willingness to meet the world on its terms.

James Lovelock expands our ideas of what is "alive" in the solar-system context, arguing that the planet itself—Gaia—is an evolved organism, meriting affection and care. Elisabet Sahtouris has written a less technical and more thoughtful appreciation of Gaia, challenging the idea that people can manage the planet. E.F. Schumacher also questions the grandiose schemes by which we would take charge of the Earth; Small is Beautiful because big is unmanageable. Aldo Leopold foresees human salvation through an extension of the ethical sense, valuing land not as a mere commodity but as a community to which we belong. Wes Jackson urges an agriculture that meets the expectations of the land before the expectations of people. Wendell Berry applies the same perspective to North American agriculture with its misplaced goals of efficiency and profits rather than of land health and cultural wholeness. George Melnyk champions bioregionalism, the development of a culture appropriate to the prairie and its traditions, a Métis culture melding the best brought to the land by the native and the immigrant.

In a philosophical vein, Lao Tsu, the ancient Chinese philosopher of Taoism, anticipated ecology in recognizing under the facade of "the ten thousand things" the interdependencies of all nature, by that recognition rejecting homocentrism and ideas of a personalized God. To seek the Tao is not to be inactive but to discern the Way of the universe and virtuously to go with the flow. Robinson Jeffers, an immersed-in-nature poet, urges love of the whole, not man apart. Aldous Huxley distills the essence of the great world religions as apprehension of the larger ineffable reality in which our egoism dissolves.

Constance Mungall and Digby McLaren edit an up-to-date compendium on the stresses and strains afflicting planet Earth. The various contributing authors provide diagnoses and prescriptions that are a mix of optimism and pessimism. The Report of the World Commission on Environment (Brundtland Report) gives a valuable overview of the state of the world environmentally and economically, though its forced optimism about environmental salvation through economic "sustainable development" detracts from its many good recommendations for environmental care and international co-operation. The public is preoccupied with saving and preserving species, and Monte Hummel's book reminds us that the focus on endangered species rather than on endangered spaces is wrong because no organisms exist apart from the land/water ecosystems in which they are embedded. Most practical, Jeremy Rifkin has edited a useful "how to" handbook on what we can do, individually and collectively, to heal the Earth.

Berry, Wendell. *The Unsettling of America: Culture and Agriculture.* New York: Avon Books, 1997.

Capra, Fritjof. *The Turning Point: Science, Society and the Rising Culture.* New York: Simon and Schuster, 1982.

Ehrenfeld, David. *The Arrogance of Humanism.* New York: Oxford University Press, 1982.

Evernden, Neil. *The Natural Alien: Humankind and the Environment.* Toronto: University of Toronto Press, 1985.

Hummel, Monte (ed). *Endangered Spaces.* Toronto: Key Porter, 1989.

Huxley, Aldous. *The Perennial Philosophy.* New York and London: Harper & Brothers, 1945.

Jackson, Wes. *Altars of Unhewn Stone: Science and the Earth*. San Francisco: North Point Press, 1987.

Jackson, Wes, Wendell Berry, and Bruce Colman (eds). *Meeting the Expectations of the Land: Essays in Sustainable Agriculture and Stewardship*. San Francisco: North Point Press, 1984.

Jeffers, Robinson. *Not Man Apart: Lines from Robinson Jeffers*. Edited by David Brower. San Francisco: Sierra Club Ballantine Books, 1969.

Lao Tsu. *Tao Te Ching: A New Translation by Gia-Fu Feng and Jane English*. New York: Vintage Books, 1974.

Leopold, Aldo. *A Sand County Almanac*. New York: Ballantine Books, 1966.

Livingston, John A. *The Fallacy of Wildlife Conservation*. Toronto: McClelland & Stewart, 1981.

Lovelock, James. *The Ages of Gaia: A Biography of Our Living Earth*. New York: W.W. Norton and Co., 1988.

Meeker, Joseph W. *The Comedy of Survival: In Search of an Environmental Ethic*. Los Angeles: Guild of Tutors Press, 1980.

Melnyk, George. *Radical Regionalism*. Edmonton: NeWest Press, 1981.

Merchant, Carolyn. *The Death of Nature: Women, Ecology and the Scientific Revolution*. San Francisco: Harper & Row, 1980.

Mungall, Constance and Digby McLaren (eds). *Planet Under Stress: The Challenges of Global Change*. The Royal Society of Canada. Toronto: Oxford University Press, 1980.

Rifkin, Jeremy (ed). *The Green Lifestyle Handbook: 1001 Ways You Can Heal The Earth*. New York: Henry Hold & Company, An Owl Book.

Sahtouris, Elisabet. *Gaia: The Human Journey from Chaos to Cosmos*. New York: Simon and Schuster, Pocket Books, 1988.

Schumacher, E.F. *Small is Beautiful*. London: Blond & Briggs, 1973.

Watts, Alan W. *The Book—On the Taboo Against Knowing Who You Are*. New York: Collier Books, 1966.

World Commission on Environment and Development. *Our Common Future*. Toronto: Oxford University Press, 1987.

NOTES

The Relict Grassland

[1] William J. Cody, *Plants of the Riding Mountain National Park, Manitoba.* Agriculture Canada Research Branch Publ. 1818/E (Ottawa: Supply and Services Canada, 1988).

The First 100 Years

[1] Carle C. Zimmerman and Garry W, Moneo, *The Prairie Community System.* Agricultural Economics Research Council of Canada (Ottawa: Agriculture Canada, 1971).

[2] E.J. Tyler, "The Farmer as a Social Class," in *Rural Canada in Transition,* ed. Marc-Adelard Tremblay and W.J. Anderson, Agricultural Economics Research Council of Canada (Ottawa: Agriculture Canada, 1966).

[3] Government of Manitoba, *Commission Report on Manitoba's Economic Future,* 1961.

[4] J.S. Shields, H.P.W. Rostad and J.S. Clayton, *Inventory of Saskatchewan Soils and their Capability for Agricultural Use,* A.R.D.A. Publ. M13, Saskatchewan Institute of Pedology (Saskatoon: University of Saskatchewan, 1970).

[5] Dennis Jones, "West's Agricultural Potential Left Undeveloped," *Saskatoon Star-Phoenix,* 30 March 1982.

[6] R.T. Coupland, *A Preliminary Ecological Analysis of Land Use in the Canadian Agricultural System* (Ottawa: Environment Canada, 1981).

[7] J.S. Rowe, "Status of the Aspen Parkland in the Prairie Provinces," in *Endangered Species in the Prairie Provinces,* ed. G. Holyroyd (Edmonton: Provincial Museum of Alberta, 1986), 27-33.

[8] Neil Evemden, "Beauty and nothingness: prairie as failed resource," *Landscape* 27(3) (1983): 1-8.

[9] J.S. Rowe and R.T. Coupland, "Vegetation of the Canadian Plains," *Prairie Forum* 9 (1984): 231-248.

The Quintessential Westerner

[1] Grant MacEwan, *Entrusted To My Care* (Saskatoon: Western Producer Prairie Books, 1986).

[2] Gene Allen, "Three Provinces Agree to Cooperate on Labrador Hydro Development," *The Globe and Mail,* 23 August 1989.

Ecology and Popular Science

[1] A.G. Tansley, "The use and abuse of vegetational concepts and terms," *Ecology* 16 (1935): 284-307.

[2] E. Warming, *Oecology of Plants: an Introduction to the Study of Plant Communities* (Oxford: Oxford Press, 1909).

[3] Carolyn Merchant, *The Death of Nature* (San Francisco: Harper & Row, 1980).

[4] R.G. Collingwood, *The Idea of Nature* (New York: Oxford University Press, A Galaxy Book, 1960).

5. Richard Levins and Richard Lewontin, *The Dialectical Biologist* (Cambridge: Harvard University Press, 1985).

[6] Peter B. Medawar, *The Art of the Soluble* (London: Methuen, 1967).

[7] John Lenihan, "Is Man a Machine?" in *The Environment and Man,* ed. John Lenihan and William W. Fletcher. *The Biological Environment, Vol. 9* (New York: Academic Press, 1979): 78-82.

[8] Jeremy Rifkin, *Declaration of a Heretic* (Boston: Routledge & Kegan Paul, 1985).

Changing the Global Vision

[1] Henri Frankfort, H. A. Frankfort, J. S. Wilson and T. Jacobsen, *Before Philosophy, the Intellectual Adventure of Ancient Man* (Harmondsworth: Penguin Books, 1949).

[2] Marshall McLuhan, *Understanding Media: The Extensions of Man* (New York: McGraw-Hill, 1964).

[3] Neil Evernden, "Seeing and being seen," *Soundings* LXVIII (1) (1985): 72-87.

[4] Maurice Merleau-Ponty, "Critique of Science," in *Philosophical Issues: a Contemporary Introduction,* ed. James Rachels and Frank A. Tillman (New York: Harper & Row, 1972): 371-73.

[5] Aldous Huxley, *The Perennial Philosophy* (New York: Harper & Brothers, 1945).

Technology and Ecology

[1] Wes Jackson, *Altars of Unhewn Stone: Science and the Earth* (San Fransisco: North Point Press, 1987).

[2] Nicholas Georgescu-Roegen, "Energy and economic myths, Part 2," *The Ecologist* 5(7): 52.

Nature, Self, and Art

[1] Walter Savage Landor, "Dying Speech of an Old philosopher." *The Norton Anthology of Poetry*, Shorter Edition (New York: W.H. Norton & Co. Inc., 1970).

[2] Elaine de Kooning, "de Kooning Memories," *Vogue* 173 (12) (1983): 394.

[3] Aldous Huxley, *The Perennial Philosophy* (New York: Harper & Brothers, 1945).

[4] Northrop Frye, *The Educated Imagination* (Toronto: Canadian Broadcasting Corporation, 1963).

[5] Friederich Nietzsche, *Beyond Good and Evil* (New York: Random House, Vintage Books, 1966): section 259.

[6] C.S. Lewis, *The Abolition of Man or Reflections on an Education with Special Reference to the Teaching of English in the Upper Forms of Schools* (London: The Centennial Press, Geoferey Bles, 1946).

[7] Paul Shepard. *Nature and Madness* (San Francisco: Sierra Club Books, 1982).

[8] Joseph W. Meeker, *The Comedy of Survival* (Los Angeles: Guild of Tutors Press, 1980).

[9] W.B. McGill, "Foreword," Agriculture & Forestry Bulletin 5 (4) (Edmonton: University of Alberta, 1982).

[10] John A. Livingston, *The Fallacy of Wildlife Conservation* (Toronto: McClelland and Stewart, 1981).

[11] Alasdair MacIntyre, *After Virtue* (Notre Dame: University of Notre Dame Press, 1981): 244-45.

[12] Mircea Eliade, *The Forbidden Forest,* trans. MacLinscott Ricketts and Mary Park Stevenson (Notre Dame: University of Notre Dame Press, 1978).

[13] Thomas Mann, *Confessions of Felix Krull: Confidence Man,* trans. Denver Lindley (New York: Knopf, 1955).

Growing Up in Granum

[1] Jose Ortegay Gasset, *Meditations on Hunting* (New York: Scribner's, 1972).

[2] Paul Shepard, *The Tender Carnivore and the Sacred Game.* (New York: Scribner's, 1973).

[3] Thorstein Veblen, *The Theory of the Leisure Class* (New York: The Modern Library, 1934).

Beauty and the Botanist

[1] William Blake, "Auguries of Innocence," in *Immortal Poems of the English Language,* ed. Oscar Williams (New York: Pocket Books, Cardinal Edition, 1952): 227-30.

[2] Galileo Galilei, *Il Saggiatore. Dialogue on the Great World Systems,* ed. Giorgio de Santillana (Chicago: University of Chicago Press, 1953).

[3] Albert Einstein, *Ideas and Opinions* (New York: Dell Publishing, Laurel Edition, 1974): 219 and 238.

[4] Ibid.

[5] James Lovelock, *The Ages of Gaia: A Biography of our Living Earth* (New York: W.W. Norton, 1988).

[6] Kathleen Raine, "A Sense of Beauty," *Resurgence,* Issue 114 (1986): 8-12.

Lake Athabasca Sand Dunes

1 J.B. Tyrrell (ed), *Journals of Samuel Hearne and Philip Turnor between the Years 1774 and 1792* (Toronto: Champlain Society, 1934).

2 J.B. Tyrell and D.B. Dowling, "Report on the Country between Athabasca Lake and Churchill River," *Annual Report for 1895*, Geological Survey, Canada Department of Mines, Ottawa: 1897.

3 J.S. Rowe and Z.M. Abougendia, "The Lake Athabasca Sand Dunes of Saskatchewan: A Unique Area," *Musk-Ox* (1982): 1-22. A condensed review of the much larger two-volume report that treats the geology, hydrology, climatology, biology, ecology and archaeology of the dune area in detail: *The Athabasca Sand Dunes of Saskatchewan, A Multidisciplinary Study*, ed. Z.M. Abougendia and W.W. Sawchyn (Saskatoon: The Saskatchewan Research Council, 1980).

4 Hugh M. Raup and George W. Argus, "The Lake Athabasca Sand Dunes of Northern Saskatchewan and Alberta, Canada. 1. The Land and Vegetation," *Botany* No. 12 (Ottawa: National Museums of Canada, 1982).

5 Georges Erasmus, "A Native Viewpoint," in *Endangered Spaces*, ed. Monte Hummel (Toronto: Key Porter, 1989): 92-98.

Crimes Against the Ecosphere

1 Law Reform Commission of Canada, *Crimes Against The Environment* [Ottawa]: Law Reform Commission Working Paper 44, 1985.

2 Pollution is the only weapon of attack on the environment recognized by the Commission, implying that the environment is safeguarded by controlling polluters.

3 Law Reform Commission of Canada, *Our Criminal Law* [Ottawa]: Law Reform Commission Report 3, 1976.

4 Maurice Cranston, "What are human rights?" *Daedalus* 12 (4) (1983): 56-59.

5 James Lovelock, *The Ages of Gaia*. (New York: W.W. Norton, 1988).

6 J.S. Rowe, "The level-of-integration concept and ecology," *Ecology* 42 (1961): 420-427.

7 International Union for Conservation of Nature and Natural Resources, *World Conservation Strategy* [Gland, Switzerland]: International Union for the Conservation of Nature, 1980.

8 C.D. Stone, "Should trees have standing?—Toward legal rights for natural objects." *Southern California Law Review* 45 (2) (1972): 450-501.

9 L.H. Tribe. "Ways not to think about plastic trees: New foundations for

environmental law." *Yale Law Journal* 83 (1974): 1315-1348.

[10] D.P Emond, "Co-operation in nature: a new foundation for environmental law." *Osgoode Hall Law Journal* 22 (1984): 323-348.

[11] George Grant, "The University Curriculum," in *The University Game,* ed. H. Adelman & D. Lee (Toronto: Anansi, 1968), 47-68.

[12] *Our Common Future,* World Commission on Environment and Development, *Annexe* 1 (Oxford: Oxford University Press, 1987).

[13] Experts Group on Environmental Law, the World Commission on Environment and Development, "All human beings have the fundamental right to an environment adequate for their health and well-being." Commentary on Article 1, *Legal Principles For Environmental Protection And Sustainable Development* (Dordrecht: Martinus Nijhoff Publishers, 1987).

Role of the University

[1] Hannah Arendt, *Thinking and Moral Consideration,* 1971. Quoted in *Manas* 31(39) (1983). See also Jose Ortega y Gasset, "The Historical Origin of Philosophy," *The Idea of Principle in Leibnitz and the Evolution of Deductive Theory* (New York: W.W. Norton, 1971), and Ortega's *Mission of the University* (Princeton: Princeton University Press, 1944).

[2] James Downey, "The University as Court Jester." *University Affairs* (May 1983).

[3] Alfred North Whitehead, *The Aims of Education* (New York: Mentor Books, 1949).

[4] Sterling M. McMurrin, *On the Meaning of the University* (Salt Lake City: University of Utah Press, 1980).

[5] *The Mission of the University of British Columbia* (Vancouver: University of British Columbia, 1979).

[6] *Preamble of the Collective Agreement* (Saskatoon: University of Saskatchewan and University of Saskatchewan Faculty Association, 1985).

[7] T .S. Eliot, "Choruses from The Rock." Chorus 1, *Selected Poems, T.S. Eliot* (Harmondsworth: Penguin Books, 1948).

Ethical Ecosphere

[1] Aldo Leopold, "The Land Ethic." *A Sand County Almanac* (Oxford, Oxford University Press, 1966), 237-64.

[2] George Bush, quoted under headline, "West Backs Gorbachev," *Saskatoon Star Phoenix,* 5 December 1989.

[3] Jonathan Miller, "The Nature of Emotion." *States of Mind* (New York: Pantheon Books, 1983).

The New Nature

[1] James Jeans, *The Mysterious Universe, Second Edition* (Cambridge: Cambridge University Press, 1933), 137.

[2] Philip P. Hanson, "What Environmental Ethics can do for You," in *Environmental Ethics, Vol. II,* ed. Raymond Bradley and Stephen Duguid (Burnaby: Simon Fraser University, 1989), 26.

[3] Robert Ardrey, *African Genesis* (New York: Dell Publishing, A Delta Book, 1963), 315-316.

[4] Francis Bacon, "The Masculine Birth of Time." *The Philosophy of Francis Bacon,* ed. and trans. Benjamin Farrington (Liverpool: Liverpool University Press, 1964), 62.

[5] Caroline Merchant, *The Death of Nature: Women, Ecology and the Scientific Revolution* (San Francisco: Harper & Row, 1980).

[6] Allan Bloom, *The Closing of the American Mind* (Toronto: Simon & Schuster, A Touchstone Book, 1987).

[7] Richard Neuhaus, *In Defense of People: Ecology and the Seduction of Radicalism* (New York: The Macmillan Company, 1971).

[8] George Melnyk, *Radical Regionalism* (Edmonton: NeWest Press, 1981).

[9] George Erasmus, "A Native Viewpoint," in *Endangered Spaces,* ed. Monte Hummel (Toronto: Key Porter, 1989), 92-98.

[10] Terry Eagleton, quoted from an article in *Commonweal,* page 39 in Richard Neuhaus's *In Defense of People.* See reference no. 7 above.

[11] Quoted from Spinoza by Alan Drengson, "Protecting the Environment, Protecting Ourselves," in *Environmental Ethics, Vol. II,* ed. Raymond Bradley and Stephen Duguid (Burnaby: Simon Fraser University, 1989), 42.

[12] Wendell Berry, "Preserving Wildness," *Resurgence* 121 (1987): 5-10.

Goals for Agriculture

[1] E.F. Schumacher, "The Proper Use of Land," *Small is Beautiful* (London: Blond & Briggs, 1973).

Too Late For Eden?

[1] Paul Shepard, *Nature and Madness* (San Francisco: Sierra Club Books, 1982).

[2] Dorothy Dinnerstein, *The Mermaid and the Minotaur—Sexual Arrangements and Human Malaise* (New York: Harper & Row, Harper Colophon Books, 1976).

[3] Dana and Wes Jackson, "The Sustainable Garden," "Investigations into Perennial Polyculture" (with Marty Bender), and "A Search for the Unifying Concept for Sustainable Agriculture," in *Meeting the*

Expectations of the Land, ed. Wes Jackson, Wendell Berry and Bruce Colman (San Francisco: North Point Press, 1984).

4 Francis Bacon, Preface to *Magna Instauratio*. Quoted by Will Durant in *The Study of Philosophy* (Toronto: Doubleday, Doran & Gundy, 1927), 133.

5 Jeremy Rifkin, *Declaration of a Heretic* (Boston: Routledge & Kegan Paul, 1985).

6 Wes Jackson, Wendell Berry and Bruce Colman (eds), *Meeting the Expectation of the Land*. (San Francisco: North Point Press, 1984).

Transforming Agriculture

1 Dan Morgan, *Merchants of Grain* (Markham: Penguin Books Canada, 1980).

2 Dennis Avery, "The US Farm Dilemma." *Science* 230 (1985): 408-412.

3 *Saskatoon Star-Phoenix*, 12 January 1988.

4 Reuters, "US Insurance Companies Begin Quiet Selloff of Farms Acquired from Foreclosures," *The Globe and Mail*, 11 December 1987.

5 *Our Common Future*, World Commission on Environment and Development (Oxford: Oxford University Press, 1987).

6 Brewster Kneen, *Land to Mouth: Understanding the Food System* (Toronto: NC Press, 1989).

7 Aldo Leopold, *A Sand County Almanac* (New York: Ballantine Books, 1949).

Prairie Land and People

1 Ralph Waldo Emerson, "Hamatreya," *The Norton Anthology of Poetry* (New York: W.W. Norton , 1970).

2 E.F. Schumacher, *Small is Beautiful* (London: Blond & Briggs, 1973), 103 and 107.

3 Aldo Leopold, *A Sand County Almanac* (New York: Ballantine Books, 1966).

4 Wendell Berry, *The Unsettling of America: Culture & Agriculture* (New York: Avon Books, 1977).

5 J.E. Weaver, *North American Prairie* (Lincoln: Johnsen Publishing 1954).

6 Edward Ahenakew, *Voices of the Plains Cree* (Toronto: McClelland & Stewart, 1973).

7 May Theilgaard Watts, *Reading the Landscape* (New York: Macmillan Publishing, 1964).

8 Stephen J. Pyne, "These Conflagrated Prairies: A Cultural Fire History of the Grasslands," in *The Prairie: Past, Present and Future*, ed. G.K. Clambey and R.H. Pemble. (Fargo, North Dakota/Moorhead, Minnesota: Proc. 9th North American Prairie Conference, Tri-College University Center for Environmental Studies, 1986), 131-137.

[9] Shakespeare's "Hamlet" Act 2, Scene 2.

[10] Wes Jackson, *Altars of Unhewn Stone: Science and the Earth* (San Francisco: North Point Press, 1987).

June Trip to China, 1983

[1] Central Intelligence Agency, *People's Republic of China Atlas*, 1971, Prepared by Superintendent of Documents, United States Printing Office, Washington, DC, 53.

China Revisited, June 1985

[1] Lao Tsu, *Tao Te Ching: A New Translation by Gia-Fu Feng and Jane English* (New York: Random House, Vintage Books, 1974), verse 1.

Modesty in the Home Place

[1] Jonathan Miller, *States of Mind* (New York: Pantheon Books, 1983), 10, 27-28.

[2] Carlos Castaneda, *Journey to Ixtlan*. (New York: Simon and Schuster, 1972).

[3] Peter Behie, "The Planetary Trust and Biocentrism: Humanity in and out of Nature, and Environmentalism in and out of Liberalism" (n.p.).

[4] Murray Bookchin, "Freedom and Necessity in Nature," *Alternatives* (13) (1986): 29-38.

[5] J.S. Rowe, "The Level-of-integration Concept and Ecology," *Ecology* 42 (1961): 420-427.

[6] E.F. Schumacher, 1973 "The Proper Use of Land," *Small is Beautiful* (London: Blond & Briggs, 1973).

STAN ROWE is Emeritus Professor of Ecology at the University of Saskatchewan. Inspired by living and working in the western grasslands and forests, he has long been a strong voice for the preservation of natural areas and for the ecocentrist movement. A prominent Canadian ecologist, Rowe won the J.B. Harkin Conservation Award in 1994 for his significant contribution to protecting Canada's parks and wilderness areas. The Canadian Botanical Association established an award in his name, honouring Rowe's contribution as the first Chair of the Ecological Section of the Association. Rowe is the author of numerous articles, reviews, and essays. In 1990 he retired to New Denver, British Columbia where he is working on writing his next collection of essays, and remains active in promoting his wholistic world view.

The Canadian Parks and Wilderness Society is Canada's grassroots voice for wilderness. Our focus is to protect wild ecosystems by establishing new parks and making sure that the needs of nature come first in their management. CPAWS has eleven chapters and 20,000 members across Canada.

For more information contact:

Canadian Parks and Wilderness Society
506 – 880 Wellington Street
Ottawa, Ontario, K1N 6K7
T: (613) 569-7226
F: (613) 569-7098
E: info@cpaws.org
www.cpaws.org

NEW LEAF PAPER

ENVIRONMENTAL BENEFITS STATEMENT

Home Place is printed on New Leaf EcoBook 100, made with 100% post-consumer waste, processed chlorine free. By using this environmentally friendly paper, NeWest Press saved the following resources:

trees	water	electricity	solid waste	greenhouse gases
9 fully grown	933 gallons	1,217 kilowatt hours	2 cubic yards	1,541 pounds

Calculated based on research done by the Environmental Defense and other members of the Paper Task Force.